T0205428

Studies in Computational Intelligence

Volume 781

Series editor

Janusz Kacprzyk, Polish Academy of Sciences, Warsaw, Poland
e-mail: kacprzyk@ibspan.waw.pl

The series "Studies in Computational Intelligence" (SCI) publishes new developments and advances in the various areas of computational intelligence—quickly and with a high quality. The intent is to cover the theory, applications, and design methods of computational intelligence, as embedded in the fields of engineering, computer science, physics and life sciences, as well as the methodologies behind them. The series contains monographs, lecture notes and edited volumes in computational intelligence spanning the areas of neural networks, connectionist systems, genetic algorithms, evolutionary computation, artificial intelligence, cellular automata, self-organizing systems, soft computing, fuzzy systems, and hybrid intelligent systems. Of particular value to both the contributors and the readership are the short publication timeframe and the world-wide distribution, which enable both wide and rapid dissemination of research output.

More information about this series at http://www.springer.com/series/7092

Jan Kozak

Decision Tree and Ensemble Learning Based on Ant Colony Optimization

 Springer

Jan Kozak
Faculty of Informatics and Communication,
  Department of Knowledge Engineering
University of Economics in Katowice
Katowice, Poland

ISSN 1860-949X   ISSN 1860-9503   (electronic)
Studies in Computational Intelligence
ISBN 978-3-030-06716-8   ISBN 978-3-319-93752-6   (eBook)
https://doi.org/10.1007/978-3-319-93752-6

Printed on acid-free paper

This Springer imprint is published by the registered company Springer International Publishing AG
part of Springer Nature
The registered company address is: Gewerbestrasse 11, 6330 Cham, Switzerland

My greatest, heartfelt thanks go to my beloved wife Dorota, who stayed with me in good and bad moments—and to my dear daughters: Asia, who showed me, that there is no such thing as impossible and that I should never give up, and Ola, who taught me how much we can achieve in life by helping others.

Thank you...

Preface

This book, as its title suggests, is devoted to decision trees and ensemble learning based on ant colony optimization. Accordingly, it not only concerns the afore-mentioned important topics in the area of machine learning and combinatorial optimization, but combines them into one. It is this combination that was decisive for choosing the material to be included in the book and determining its order of presentation.

Decision trees are a popular method of classification as well as of knowledge representation. As conceptually straightforward, they are very appealing and have proved very useful due to their computational efficiency and generally acceptable effectiveness. At the same time, they are easy to implement as the building blocks of an ensemble of classifiers—in particular, decision forests. It is worth mentioning that decision forests are now considered to be the best available off-the-shelf classifiers. Admittedly, however, the task of constructing a decision tree is a very complex process if one aims at the optimality, or near-optimality, of the tree. Indeed, we usually agree to proceed in a heuristic way, making multiple decisions: one at each stage of the multistage tree construction procedure, but each of them obtained independently of any other. However, construction of a minimal, optimal decision tree is an NP-complete problem. The underlying reason is that local decisions are in fact interdependent and cannot be found in the way suggested earlier.

The good results typically achieved by the ant colony optimization algorithms when dealing with combinatorial optimization problems suggest the possibility of using that approach effectively also for constructing decision trees. The underlying rationale is that both problem classes can be presented as graphs. This fact leads to the possibility of considering a larger spectrum of solutions than those based on the heuristic mentioned earlier. Namely, when searching for a solution, ant colony optimization algorithms allow for using additional knowledge acquired from the pheromone trail values produced by such algorithms. Moreover, ant colony opti-mization algorithms can be used to advantage when building ensembles of classifiers.

This book is a combination of a research monograph and a textbook. It can be used in graduate courses, but should also be of interest to researchers, both specialists in machine learning and those applying machine learning methods to cope with problems from any field of R&D. The topics included in the book are discussed thoroughly. Each of them is introduced in such a way as to make it accessible to a computer science or engineering graduate student. All methods included are discussed from many angles, in particular regarding their applicability and the ways to choose their parameters. In addition, many unpublished results are presented.

The book is divided into two parts, preceded by an introduction to machine learning and swarm intelligence. The first part discusses decision tree learning based on ant colony optimization. It includes an introduction to evolutionary computing techniques for data mining (Chap. 2). Chapter 3 describes the most popular ant colony algorithms used for learning decision trees. Chapter 4 introduces some modifications of the ant colony decision tree algorithm. Examples of practical applications of the methods discussed earlier are presented in Chap. 5.

The second part of the book discusses ensemble learning based on ant colony optimization. It begins (Chap. 6) with an introduction presenting known solutions of a similar type—more precisely, evolutionary computing techniques in ensemble learning—as well as a literature review. Chapter 7 includes formal definitions and example applications of ensembles based on random forests. Chapter 8 introduces an adaptive approach to building a decision forest with ant colony optimization. The book concludes with some final remarks and suggestions for future research.

Let me end with mentioning the persons whose advice, help and support were indispensable for writing this book. Special thanks are due to Prof. Jacek Koronacki for the possibility of collaboration, his extraordinary commitment, fruitful discussions and all comments. Similar thanks to Prof. Beata Konikowska, who showed unusual meticulousness and gave me many valuable remarks and pieces of advice. I would also like to thank Prof. Mikhail Moshkov for his continuing comments and long discussions. Sincere thanks are due to Dr. Przemysaw Juszczuk, who has been deeply involved in creating this book.

Katowice, Poland   Jan Kozak
October 2017

Contents

Chapter 1
Theoretical Framework

1.1 Classification Problem in Machine Learning

One of the main problems in the machine learning field is classification. An important method used in classification problems is the prediction model known as decision tree. This is a popular classifier, which can be represented as a graph. The latter feature is important, especially from the viewpoint of algorithms supporting the problem solving process.

A classical decision tree construction is carried out with the use of statistical methods. However, in view of the graph representation of decision trees and the large space of possible solutions, other methods can be employed as well. One of the methods used in the swarm intelligence field (as a branch of artificial intelligence) is ant colony optimization (ACO), which seems to yield good results for problems that can be presented as graphs.

These observations (presented in Fig. 1.1) have been the crucial inspiration for this book. They also have a consequence in the form of analysis of the possibilities of using ACO to construct an ensemble of classifiers based on decision trees (a decision forest). During a single run, ACO generates a set of potential solutions, part of which are eventually eliminated with the use of some dedicated mechanisms.

Development of an ensemble from carefully selected solutions allows us to create some kind of consensus between them—specific for the decision forest based on ant colony optimization. The most important point of this book is the application of ant colony optimization to the problem of learning decision trees and ensemble methods.

Machine learning can be seen as a process where a machine "learns" to solve a problem Q on the basis of empirical experience W and a quality measure Y. Together with the growth of experience W, the quality Y of the solution should also increase [42]. Thus machine learning can be seen as a research field within which methods for dealing with data mining problems are built.

Historically, machine learning has been divided into unsupervised and supervised learning. One of the branches of supervised learning is supervised classification (or, for short, classification)—and in this book we will focus on this branch only.

© Springer International Publishing AG, part of Springer Nature 2019
J. Kozak, *Decision Tree and Ensemble Learning Based on Ant
Colony Optimization*, Studies in Computational Intelligence 781,
https://doi.org/10.1007/978-3-319-93752-6_1

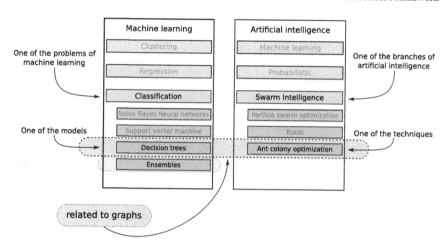

Fig. 1.1 Problem positioning

In supervised classification, we are given a set of samples (also called a training set). This set consists of n observations (also called objects or samples):

$$X = \{x_1, x_2, \ldots, x_i, \ldots, x_n\}, \tag{1.1}$$

Each of the observations x_i is described by m attributes (also called features)

$$a_1, a_2, \ldots, a_m, \tag{1.2}$$

with $a_j \in A_j$, $j = 1, \ldots, m$, where A_j denotes the domain of the j-th attribute. In this way, features $a_1, a_2, \ldots, a_m$ form a feature space $A_1 \times A_2 \times A_m$.

The values of these attributes can be quantitative (for example, price) or categorical (for example, weather: "rainy" or "sunny"). Each observation belongs to one of C different (finitely many) and known decision classes. Therefore, each observation can be represented as:

$$x_i = (\mathbf{V_i}, c_i), v_i^j \in A_j, c_i \in \{1, \ldots, C\}, \tag{1.3}$$

where $\mathbf{V_i} = [v_i^1, \ldots, v_i^m]$ is a vector in an m-dimensional feature space, v_i^j is the value of attribute a_j for observation (object) x_i, and c_i is the class label (also called the decision class) of that observation (object x_i).

Accordingly, X can be represented as

$$X : \{(\mathbf{V_i}, c_i)\}_{i=1}^n. \tag{1.4}$$

Based on the definitions given above, a classification problem can be defined as determining how to assign an object to a class, knowing that there are C different decision classes and that each object belongs to one of them. A learning algorithm $\mathcal{L}$ is first trained on a set of pre-classified examples X. In the classification, each c_i takes one of the C nominal values. X consists of independently and identically distributed samples obtained according to a fixed, but unknown, joint probability distribution κ_{c_i} in the feature space in each class.

The goal of classification is to produce a classifier which can be used to divide a set of objects into distinct classes (classification of objects), and, furthermore, to evaluate the classification performed. Thus we can say that in this process a hypothesis h is proposed that approximates the evaluation function F best (with respect to a selected classification quality measure, for example: accuracy, precision or recall). That is, h minimizes the loss function (i.e., zero-one loss) in the space of feature vectors and classes $\mathbf{V} \times C$ based on distribution κ_{c_i}.

Classification begins when the learning algorithm $\mathcal{L}$ receives as its input a training set X and conducts search over a hypothesis space $H_{\mathcal{L}}$ that approximates the true function F. More exactly, the learning algorithm is a mapping $\mathcal{L} : S \rightarrow H_{\mathcal{L}}$, where S is the space of all training sets of size n, which maps the training set to a hypothesis. The selected hypothesis h can then be used to predict the class of unseen examples [29].

It should be mentioned that classification, in its classical approach, differs from the so-called reinforcement learning. Within the latter approach, the learning algorithm includes some feedback procedures that directly inform the algorithm about the achieved quality. That learning method allows us to find solutions without the need to use the knowledge acquired earlier.

In the classical approach, classifiers are generated on the basis of the acquired knowledge (training set) without any additional feedback. However, in case of classifiers built using stochastic methods (like ant colony optimization algorithms), the situation is different. In such a case, agents (to be more specific, the ant agents, which are described in detail in Sect. 1.3) acquire information on the quality of the current solution.

1.2 Introduction to Decision Trees

One of popular data classification methods are decision trees. Their application is particularly important due to the simple and effective (in terms of computational complexity) object classification process. They allow for simplifying the process of building an ensemble of classifiers: for example, decision forests, which are defined as sets of specific decision trees. This allows us to improve the quality of classification. The process of constructing decision trees is a complex problem of multi-criterial decisions concerning the rules used in the optimal data split.

Decision trees may easily represent complex concepts whose definitions can be described based on the feature space. In other words, trees can represent a function which maps attribute values (or values of the features) into a set of decision classes (or, in other words, class labels) which represent a permissible hypothesis [38].

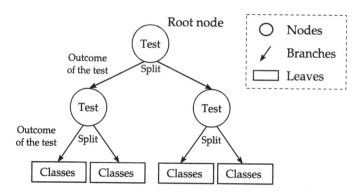

Fig. 1.2 Binary decision tree

Moreover, compared to other algorithms, decision trees are very competitive with respect to memory usage. In addition, a decision tree is a classifier which produces decision rules. Decision trees are commonly used in operational research, specifically in decision analysis, to identify the optimum strategy for reaching a goal.

A decision tree is an acyclic, directed graph in which all vertices are called nodes, edges are called branches, vertices without descendants—leaves, and the root is the only vertex without a parent. All nodes contain tests on the attributes A, which are generated according to the selected splitting criteria. They represent the way to determine the data split (described in detail further in this chapter) which divides data according to the attribute values for that data. Finally, all test results are represented by branches. An example tree structure is presented in Fig. 1.2.

A special case of decision trees are binary decision trees. In such trees, every node (except the terminal ones—leaves) has exactly two descendants. Thus objects (observations) are each time divided in two subsets. Decision trees of this type are built, for example, with the CART algorithm (described in detail below), or one of the algorithms crucial from the viewpoint of this book—ant colony decision trees, described in Sect. 3.2.

1.2.1 Example

The majority of examples as well as their solutions presented in this book are based on learning decision trees or ensemble methods on the basis of knowledge written down as decision tables. Decision tables are a simple form of presenting the acquired observations and information which will be used in the algorithm learning process. The recorded information does not determine the way of processing the decision, the decision recipient, and the input and output data. The decision tables serve mostly as data storage, and are also used for further verification of the classifier quality.

Table 1.1 Decision table

	Attributes				Decision attribute
	a_1	a_2	a_3	a_4	c
x_1	1	0	0	1	1
x_2	1	0	1	1	0
x_3	0	0	0	1	1
x_4	0	0	1	0	0
x_5	1	1	0	0	0
x_6	0	1	0	0	1

Thus we can say that a decision table, written down as an ordered pair (Eq. (1.5)), represents the problem, for which the classifier is being built.

An example decision table (which will be analyzed in Sect. 3.1) can also be presented like Table 1.1, where the set of conditional attributes and the decision attribute are given. The observations are presented as objects $x_1 \ldots x_6$.

$$(X, A \cup \{c\}), \tag{1.5}$$

where X is the set of objects, and A is the set of attributes, including the decision attribute—c.

To better understand decision trees, let us give a simple example of a binary decision tree construction process. The idea behind the whole process is very straightforward. The initial node (root) takes into account the whole set X. Then the learning set X is split in two subsets X_1 and X_2 with $X_1 \cup X_2 = X$, and each of those subsets is assigned to a different node. In the subsequent steps, the whole process is repeated in such a way that X_1 is divided into $X_{1,1}$ and $X_{1,2}$, where $X_{1,1} \cup X_{1,2} = X_1$, while X_2 is divided into $X_{2,1}$ and $X_{2,2}$, where $X_{2,1} \cup X_{2,2} = X_2$, and so on.

This procedure is repeated as long as the stop criterion is not met—for example, while objects from different decision classes can still be found in the same subset (thus until the time when the maximal separation of objects is achieved). Then such a subset is often called a terminal subset, and is marked as a leaf of the decision tree. With such an assumption, the process of building the decision tree for the decision table presented as Table 1.1 would last as long as any of the objects x_2, x_4, x_5 is found in the same subset with any of the objects x_1, x_3, x_6.

As we can see, the crucial problem is how the process of splitting the set X in two subsets (and further splits of those subsets as well) should be carried out. The detailed solutions used in popular algorithms are presented further on in this book. However, generally speaking, we can say that subsequent divisions of the set should lead to increasing the homogeneity of objects in the resulting subsets, and so to increasing the "purity" of those subsets, or, in other words, decreasing the entropy.

Fig. 1.3 Example of a
decision tree including four
decision classes

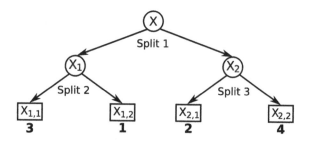

To give a brief description of this issue, let us take a look at a problem with 4 decision classes uniformly distributed in the set X, which can be divided into subsets corresponding to those decision classes: $X_{c=1}$, $X_{c=2}$, $X_{c=3}$ and $X_{c=4}$, respectively. The described process is presented in Fig. 1.4.

Assume each of the decision classes contains two objects. Then the distribution can be represented by probabilities p_1, p_2, p_3, p_4, where p_i, $i = 1, \ldots, 4$, denotes the probability of an object belonging to the i-th decision class occurring in the given node, and can be written in the form $\frac{|X_{c=i}|}{|X|}$.

Thus the distribution is as follows: $(\frac{1}{4}, \frac{1}{4}, \frac{1}{4}, \frac{1}{4})$. As a result of the division we can, for example, obtain subsets with the distribution of affiliation to the decision classes as follows: $(0, \frac{1}{4}, \frac{1}{2}, \frac{1}{4})$ and $(\frac{1}{2}, \frac{1}{4}, 0, \frac{1}{4})$. This would mean that the first subset (node of the tree) does not include any of the objects belonging to the first class; however, it includes one object from the second class, both objects from the third class, and one object from the fourth class. Analogically in case of the second subset, as can be clearly seen in Fig. 1.4a. We can also obtain the division $(0, 0, \frac{1}{2}, \frac{1}{2})$ and $(\frac{1}{2}, \frac{1}{2}, 0, 0)$, which would mean that for each subset we only have objects representing two decision classes (as can be seen in Fig. 1.4b).

In the general sense, such a division seems potentially better. However, this is an obvious simplification, and the description of constructing a binary decision tree is presented further [31, 44]. For the situation presented in such a manner, we can say that the simple decision tree presented in Fig. 1.4c would be built.

In case of a more complex problem (involving more attributes), construction of a decision tree based on Table 1.1 can be carried out based on divisions for each attribute (for simplification, in this particular example they are analyzed separately). In such a situation, if the division at the first node t is made based on attribute a_1, the left subtree t_l would contain a set of objects containing, for example, 0 on the position of attribute a_1: x_3, x_4, x_6, while the right subtree t_p would contain the set of objects x_1, x_2, x_5 (see Fig. 1.5). Therefore, the initial distribution of affiliations to the decision classes $(\frac{1}{2}, \frac{1}{2})$ at the subsequent nodes would be as follows: $(\frac{1}{3}, \frac{2}{3})$ in t_l and $(\frac{2}{3}, \frac{1}{3})$ in t_p. Similar divisions could be carried out on the basis of other attributes.

Afterwards, the resulting subsets would be divided further until the maximum uniformity of decision classes at the terminal nodes is achieved. In case of t_l, this could be e.g. the division based on the values of attribute a_3. Such division would allow us to obtain a subsequent node $t_{l,l}$ containing the subset of objects representing decision class 0, i.e. the set x_3, x_6. These are objects with descriptors $(a_1, 0)$ and

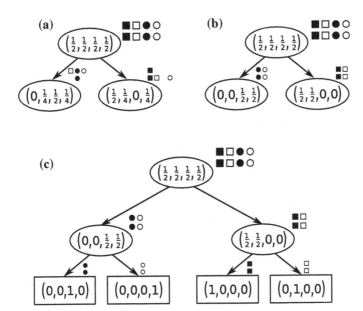

Fig. 1.4 Example of a decision tree with a given distribution of affiliation to decision classes. **a** Example division at the first node. **b** Other example division at the first node, **c** whole decision tree presented as continuation of figure (**b**)

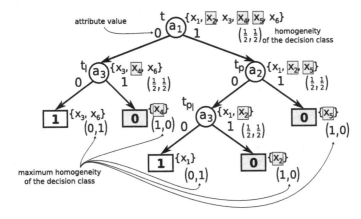

Fig. 1.5 Example process of building the decision tree with a given distribution of affiliations to decision classes based on Table 1.1

$(a_3, 0)$—i.e., objects for which the values of both attribute a_1 and attribute a_3 equal 0. An analogous situation would be observed at the subsequent nodes, as presented in detail in Fig. 1.5.

1.2.2   Decision Tree Construction

Two very well known algorithms for constructing decision trees are: C4.5 (and earlier ID3), where in this particular case attribute tests are carried out at every node, and CART, where all tests are carried out for all combinations of attributes and their values.

The ID3 [48] algorithm constructs a decision tree in a "top-down" fashion. ID3 has proven very useful, but there are multiple restrictions that make it not applicable in many real-world situations. In turn, C4.5 was developed to deal with these problems, and can be considered a good solution for low-quality or missing data, continuous variables, or large-size data [50].

The classification and regression tree (CART) approach was developed by Breiman et al. in 1984 [5]. CART is characterized by the fact that it constructs binary trees. As the heuristic function of ACDT is based on CART, the latter algorithm is most relevant to our work. Splits are selected using the Twoing and Gini criteria. CART looks for splits that minimize the squared prediction error. The decision tree is built in accordance with a splitting rule that performs multiple splitting of the learning sample into smaller parts.

Usually, a "divide and conquer" strategy is used to built the decision tree through recursive partitioning of the training set. This means we should be able to divide the problem, solve it, and then compile the results. The process of building the decision tree begins with choosing an attribute (and a corresponding value of that attribute) for splitting the data set into subsets. Selection of the best splitting attribute is based on heuristic criteria.

Construction of an optimal binary decision tree is an NP–complete problem, where an optimal tree is one that minimizes the expected number of tests required for identification of unknown objects (as shown in [31]). However, we should remember that we can also optimize the depth of decision trees, the number of nodes, or the classification quality.

The problem of designing storage-efficient decision trees based on decision tables was examined in [44]. They showed that in most cases construction of a storage-optimal decision tree is an NP-complete problem, and hence a heuristic approach to that task is necessary. Construction of an optimal decision tree may be defined as an optimization problem where, at each stage of decision making, an optimal data split is selected [54].

Splitting criteria and attribute test

Splitting criteria are used to find the best attribute test—capable of dividing the data set at a given node in two (or more, depending on the decision tree type) data subsets, uniform with respect to the decision class. By a test we mean a condition used for dividing the data.

The splitting criteria at every node determine the attribute (or set of attributes) according to which all observations will be divided. Such attribute tests represent the condition applied at every node (depending on the decision tree structure, as well as on the split criterion used). Below we present different kinds of tests (for binary trees):

- equality test (for discrete and nominal attributes), which allows for splitting the data according to the values of attributes—the object and the test condition must have the same attribute value:

$$T(x) = \begin{cases} 1, & \text{if } a_i(x) = v \\ 0, & \text{in other case,} \end{cases} \quad (1.6)$$

- inequality test (for ordered and continuous attributes) which divides the data according to a given attribute threshold. If the value of an attribute is greater than the threshold, than such object passes the test:

$$T(x) = \begin{cases} 1, & \text{if } a_i(x) > cut \\ 0, & \text{in other case,} \end{cases} \quad (1.7)$$

- membership test (for discrete attributes, in specific cases for continuous attributes as well) is a generalization of the equality test, and splits the data according to a set of attribute values. An object fulfills the test condition if the value of the attribute belongs to a predefined set of values for the test:

$$T(x) = \begin{cases} 1, & \text{if } a_i(x) \in V \\ 0, & \text{in other case,} \end{cases} \quad (1.8)$$

- and others.

The notational convention used in the above equation is as follows:
x—object,
v—attribute value,
V—set of attribute values,
a_i—conditional attribute,
cut—cutoff or threshold value.

Splitting criteria in the CART algorithm

There is no doubt that selection of the data division at every single node is the hardest and most complex phase of constructing a decision tree. For example, consider the CART algorithm (this solution is convergent to ant colony decision trees). To evaluate the test, in most cases an impurity function $i(m)$ (where m denotes the current node) is specified (see (1.9)). The function enables estimation of the maximum homogeneity of the child nodes. Since the impurity function of the parent node m_p is constant for every possible division $a_j \leq a_j^R$, $j = 1, \ldots, M$ (where M denotes the number of attributes, and a_j^R is the best possible division for attribute a_j), the maximum homogeneity of the left and right descendants is determined by the maximum difference in the impurity function $\Delta i(m)$ (diversity measure) [5]:

$$\Delta i(m) = i(m_p) - P_l i(m_l) - P_r i(m_r), \tag{1.9}$$

where:

P_l—probability of the object passing to node m_l (left sub-tree),
P_r—probability of the object passing to node m_r (right sub-tree).

The algorithm for constructing the decision tree solves the maximization problem at the phase of selecting the division for every node. It searches all possible values of attributes, which allows for finding the best possible division (the highest value of the diversity measure, i.e. of the difference in the impurity function) [5]:

$$\underset{a_j \leq a_j^R, j=1,\dots,M}{\operatorname{argmax}} \left[i(m_p) - P_l i(m_l) - P_r i(m_r) \right]. \tag{1.10}$$

For the CART algorithm, [5] proposed two criteria for splitting a node—two ways to calculate the diversity measure: Gini and Twoing (split into two parts). Both the rules presented below were included (separately) in the heuristic function of ACDT [4].

The Gini splitting rule is based on the Gini index, which is a measure of random variable concentration. The overarching goal in this case is to make a split into possibly uniform cases at the descendant nodes. The impurity function is calculated using the formula:

$$i(m) = \sum_{c \neq o} p(o|m) p(c|m), \tag{1.11}$$

where:

$p(c|m)$—probability of obtaining decision class c at node m,
$p(o|m)$—probability of obtaining decision class o at node m.
The condition according to which the split is made is determined based on Eqs. (1.9) and (1.10). This yields the following formula:

$$\underset{a_j \leq a_j^R, j=1,\dots,M}{\operatorname{argmax}} \left(-\sum_{c=1}^{C} p^2(c|m_p) + P_l \sum_{c=1}^{C} p^2(c|m_l) + P_r \sum_{c=1}^{C} p^2(c|m_r) \right), \tag{1.12}$$

where:

$p(c|m_p)$—probability of obtaining decision class c at node m_p (at the current node),
$p(c|m_l)$—probability of obtaining decision class c at node m_l (in the left sub-tree),
$p(c|m_r)$—probability of obtaining decision class c at node m_r (in the right sub-tree),
C—number of decision classes.

The Twoing criterion divides the data in two possibly equal parts (two subsets). In this case, homogeneity of the decision class is less important than in case of the Gini criterion, although it still plays a certain minor role. The diversity measure in this particular case is defined as:

$$\Delta i(t) = \frac{P_l P_r}{4} \left[\sum_{c=1}^{C} |p(c|m_l) - p(c|m_r)| \right]^2. \tag{1.13}$$

The division condition is calculated on the basis of Eqs. (1.9) and (1.10), which may be written down as [5]:

$$\underset{a_j \le a_j^R, j=1,\dots,M}{\operatorname{argmax}} \left(\frac{P_l P_r}{4} \left[\sum_{c=1}^{C} |p(c|m_l) - p(c|m_r)| \right]^2 \right). \tag{1.14}$$

Splitting criteria in the C4.5 algorithm

Frequently used splitting criteria are rules based on entropy, like those known from ID3 [48]: information gain, or gain ratio. They are applied in C4.5 [50] as well as in the Ant-Miner algorithms [47] (to be discussed later). If such criteria are used, the resulting decision trees are not necessarily binary. This is due to fact that tests at the nodes correspond to attributes, and branches to possible values of those attributes (for discrete data). For every node, the division with the highest value of the relative information gain is selected:

$$\underset{a_j \le a_j^R, j=1,\dots,M}{\operatorname{argmax}} \left(\frac{information\_gain(a_i, S)}{entropy(a_i, S)} \right), \tag{1.15}$$

where $information\_gain(a_i, S)$ is the information gain from (1.16), and $entropy$ (a_i, S) is the entropy of data distribution in the set S based on the value of attribute a_i (1.17).

$$information\_gain(a_i, S) = entropy(y, S) - \sum_{c=1}^{C} \frac{|S_c|}{|S|} \cdot entropy(y, S_c) \tag{1.16}$$

$$entropy(y, S) = \sum_{j=1}^{|y|} -\frac{|S_j|}{|S|} \cdot \log_2 \frac{|S_j|}{|S|} \tag{1.17}$$

Pruning of Decision Trees

Excessive fitting of a decision tree to the cases in the training set may lead to over-training (overfitting) of the decision tree. This is the situation where conditions at the decision tree nodes depend on the data contained in the training set. Such a situation results in a very good classification for the training set, but new data, unknown so far, is classified poorly, with large errors.

Moreover, decision trees constructed in the way described above are often of a large height. To remedy this, a so-called pruning mechanism is applied—which consists most often in replacing nodes (sub-trees) with leaves. Such replacement is carried out according to a pruning criterion, which is often used in the process classifying data from an additional training set (pruning). In such a case, these are objects which did not take part in the decision tree learning process. Other pruning methods that do not require generating an additional training set can be used as well.

The most popular pruning methods that do not need (but allow for) the use of an additional pruning set are: pessimistic pruning (error-based pruning) [50] used in the C4.5 algorithm and cost complexity pruning [5] used in the CART algorithm. However, more pruning methods can be distinguished: for example, reduced error pruning [49], which requires using an additional data set, minimum error pruning [45] and others. Comparison and description of the remaining pruning methods can be found in numerous publications, for example [19, 41, 49, 54].

Another concept closely related to decision tree pruning is the stop condition used in the process of building decision trees. This is a criterion which allows for defining the point at which further construction of the tree should be stopped. Such construction involves calling a recursive function that is responsible for extending a branch beyond the given node. Thus the stopping process depends on the given path of the tree. Detailed solutions for pruning and stopping extension of a decision tree are described in Chap. 3. Moreover, Sect. 3.1 presents an example schema of building a decision tree based on the CART algorithm.

1.3 Ant Colony Optimization as Part of Swarm Intelligence

Swarm intelligence is an innovative distributed intelligent paradigm for solving optimization problems. It was originally inspired by biological examples of swarming, self-organizing foraging phenomena in social insects. There are many examples of swarm optimization techniques, such as: particle swarm optimization, artificial bee colony optimization and ant colony optimization. The last approach deals with artificial systems inspired by natural behaviors of real ants, especially foraging behaviors based on pheromone substances laid on the ground.

The fundamental concept underlying the behavior of social insects is self-organization. Swarm intelligence systems are complex—they are collections of simple agents that operate in parallel and interact locally with each other and their environment to produce the emergent behavior.

The basic characteristics of metaheuristics inspired by nature can be summarized as follows [1, 7, 40]: such heuristics

- model a natural phenomenon,
- are stochastic,
- in case of multiple agents, often have parallel structure,
- use feedback information for modifying their own parameters—i.e., are adaptive.

Development of algorithms that utilize certain analogies with nature and social insects to derive non-deterministic metaheuristics capable of yielding good results in hard combinatorial optimization problems could be a promising field of research.

Many features of the collective activities of social insects are self-organized. Self-organization theory (SO) [8] was originally developed in the context of physics and chemistry to describe the emergence of macroscopic patterns out of processes and interactions defined at the macroscopic level. It can be extended to social insects to show that complex collective behavior may emerge from interactions among individuals that exhibit simple behavior. In those cases, there is no need to refer to the individual complexity to explain complex collective behavior. Recent research shows that SO is indeed a major component of a wide range of collective phenomena in social insects [3].

Self-organization among social insects often requires interaction between insects: such interactions can be either direct or indirect. Direct interactions include the following: antennation, trophallaxis (food or liquid exchange), mandibular contact, visual contact, chemical contact (odour of nearby nestmates), etc. Indirect interactions are more subtle: two individuals interact indirectly when one of them modifies the environment, and the other responds to the new environment at a later time.

A heuristic is defined by [52] as a technique which seeks good (i.e. near-optimal) solutions at a reasonable computational cost without being able to guarantee either feasibility or optimality, or even—in many cases—to state how close to optimality a particular feasible solution is. Heuristics are often problem-specific, so that a method which works for one problem cannot be used to solve a different one.

In contrast, metaheuristics are powerful techniques, generally applicable to a large number of problems. A metaheuristic is an iterative master strategy that guides and modifies the operations of subordinate heuristics by intelligently combining different concepts for exploring and exploiting the search space [24, 46]. At each iteration, a metaheuristic may manipulate a complete (or incomplete) single solution or a collection of solutions. The success of these methods is due to the capacity of such techniques to solve some hard combinatorial problems in practice.

Over the last decades, numerous stochastic methods for problem optimization have been developed. One of the biggest optimization problems concerns the number of dimensions (which is very important in data mining). More complex problems can easily reach a few hundreds of dimensions, where each parameter may take values in a different range. In this case, stochastic methods are sometimes used. They often take inspiration from biological or social behavior. Most of stochastic techniques are based on the population of individuals which are improved in the subsequent iterations of the algorithm. Eventually, after meeting the stop criteria, the whole algorithm is stopped on the basis of a predefined measure. It is worth noting that the concept of a metaheuristic is very general.

Local search techniques are often deemed to belong to the same class as metaheuristics. The first approximate methods associated with stochastic approximation were introduced in 1951 [53]. Stochastic approximation was one of the first methods based on the use of random variables. However, it is still very popular, and is studied in depth in the newest articles [59]. In 1952, the Kiefer-Wolfowitz method was presented [35], which was considered a generalization of the stochastic approach.

The first evolutionary strategies were based on random search related to modification of a single entity. There was no population concept introduced in later nature-inspired techniques. The next and very important stage of developing nature-inspired concepts involved evolutionary programming and genetic programming [21].

Evolutionary programming is often considered one of four main evolutionary algorithm paradigms. The most important difference between the two above methods is the program structure. In evolutionary programming, the program structure is fixed, and the parameters are allowed to evolve. Both concepts are correlated with automatic creation of computer programs. Genetic programming was preceded by a genetic algorithm, introduced in 1975 [30].

One of the latest concepts connected with genetic algorithms are so-called quantum genetic algorithms [55]. A quantum genetic algorithm is a combination of a genetic algorithm and quantum computing. In quantum computing, the smallest unit of information storage is a quantum bit (qubit). In general, a quantum algorithm is less complex than its classic equivalent due to the concept of quantum superposition.

A very important concept was introduced in [36], where the simulated annealing method was described. This algorithm has strong mathematical foundations and is based on certain physical cooling processes.

Over the following years, the above concepts were extensively studied and developed. At the same time, various nature-based methods were introduced. Among them, we should mention ant colony optimization [10] (which is described below), memetic algorithms [43] and particle swarm optimization proposed in by Kennedy and Eberhart in [18].

The continuously growing interest in various nature-inspired algorithms has led to creating more and more complicated algorithms. Of crucial importance for those developments was the introduction of an artificial bee colony—a stochastic technique developed by Karaboga in 2005 [33]. One of the newest swarm intelligence algorithms is the Krill herd algorithm [23, 58]. It is worth noting that the said algorithm shows there is still growing demand for similar concepts—publications on this subject continued until 2016 [6, 57].

Some of the most popular swarm intelligence algorithms are charted in Fig. 1.6. They include one of the greatest importance from the viewpoint of this work—the ant colony optimization approach. However, if we wanted to follow this subject, a large collection of different algorithms could be indicated. A good review of popular articles on swarm and nature-inspired algorithms can be found in [20].

1.3.1 Ant Colony Optimization Approach

Ant colonies exhibit very interesting collective behavior: even though a single ant has only simple capabilities, the behavior of the whole ant colony is highly structured. This is the result of co-ordinated interactions and co-operation between the ants (agents), and represents swarm intelligence. This term is applied to any work involving the design of algorithms or distributed problem-solving devices inspired by collective behavior of social insects [3].

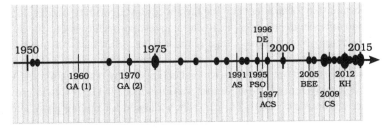

GA - Genetic algorithm
AS - Ant system
PSO - Particle swarm optimization
DE - Differential evolution
ACS - Ant colony system
BEE - Bees algorithm
CS - Cuckoo search
KH - Krill herd

Fig. 1.6 Nature inspired metaheuristics developed over the last 40 years

The main idea of ant colony algorithms is inspired by stigmergy—which may be understood as the ability of swarm members to communicate within the swarm by changing the habitat (for example, through pheromone trails in case of ants). The stigmergy phenomenon may be seen as a result of natural social behavior of the population members motivated by their main goal. To be more specific, it is the effect of cooperation with positive feedback. This phenomenon was first introduced by Grasse [25, 26].

Thus ant colony algorithms may be considered as an attempt to reflect the idea of a colony of ants cooperating to find food. The analogy of foraging is the fitness function for the problem to be solved by the algorithm—thus ants are (in the sense of the algorithm) equivalent to ant agents. Generally speaking, an ant agent can be seen as an algorithm together with the data necessary for its implementation.

Use of the mentioned ant agent allows us to solve the given optimization problem. The information on the solution is acquired from the environment (solution space) and enables the agent to progress towards a specific goal. The solution reached is based not only on the knowledge—the heuristic function, but also on the experience of the previous ant agents and of the population—which reflects the pheromone trail values, the prize for the solution selected in the preceding algorithm steps.

In any state of the system accessible to an ant agent, the agent can choose among several available actions. Accordingly, the agent has to decide which action it should perform. The main problem of a multi-agent system is to find the optimal or nearly optimal decision—or, to be more specific, a series of consecutive decisions. The agent's decision is influenced by two factors: value of the pheromone function utility (common memory) and some heuristic which depends on the problem being solved. The agent makes the decision in a completely non-deterministic way. Due to this, in case of such an algorithm there is no need to analyze conflicts among the ant agents.

An important step towards developing the ant colony algorithm was development of the Dorigo ant system [10, 13, 15]. It was a novel approach based on the analogy to a real ant colony, which gained huge popularity and allowed for achieving very good results [12]. Through various extensions, the ant system was transformed into a new approach used in optimization—so-called ant colony optimization (ACO).

The ant colony system (ACS) algorithm was introduced in [13, 14] to improve the performance of the Ant System [14, 22], which allowed for finding good solutions within a reasonable time for small size problems only. The ACS was based on three modifications to the ant system:

- different node transition rule,
- different pheromone trail updating rule,
- use of local and global pheromone updating rules (to favor exploration).

ACO was proposed by Dorigo and colleagues [17] as a method for solving static combinatorial optimization problems. ACO algorithms are part of swarm intelligence, i.e. a research field that studies algorithms inspired by observation of swarm behavior [9, 11]. Swarm intelligence algorithms use simple individuals that cooperate through self-organization, i.e., without any form of central control over the swarm members. A detailed overview of the self-organization principles exploited by those algorithms in addition to examples from nature can be found in [26].

The mechanisms behind ACO are stigmergy together with implicit solution evaluation and autocatalytic behavior. The basic idea of ACO follows biological inspiration very closely. Therefore, there are many similarities between the real ants and the ant agents (i.e., artificial ants). Both real and artificial ant colonies are composed of a population of individuals that work together to achieve a certain goal. A colony is a population of simple, independent, asynchronous agents that cooperate in order to find a good solution to the problem at hand. In case of real ants, the problem is to find food, while in case of artificial ants the problem is to find a good solution to a given optimization problem. A single ant (either a real or an artificial one) is able to find a solution to its problem, but only cooperation among many individuals through stigmergy enables the ants to find good solutions.

1.3.2 Example

Very good results achieved by ant colony optimization algorithms can be seen especially in case of problems whose structure can be presented as a graph. This is due to the ease of mapping the path traveled by an ant agent, and to the capability of storing pheromone directly on the visited edges (path). Thus a good example application for the mentioned algorithm is the problem of finding the shortest path in a weighted graph.

Obviously, a practical solution to that problem would be found with Dijkstra's algorithm, Bellman-Ford algorithm, the A* search algorithm, or other well-known method. However it is easy to present the efficiency of the ant colony optimization algorithm on the example of the mentioned problem without necessity to use it in a real application.

For the purpose of the considered example, we will treat the shortest path problem as a maximization problem. Accordingly, the main goal will be to find the longest path between two points in an acyclic, directed graph. This problem is commonly called the longest path problem. An example graph $G = (V, E)$, where V is a set of nodes (also called vertices), and A is a set of ordered pairs of nodes (also called arrows or directed edges) is presented in Fig. 1.7. Graph G consists of the 6-element set of vertices $V = \{a, b, c, d, e, f\}$ along with the 9-element set of edges $A = \{(a, b, 9), (a, c, 5), (b, d, 5), (b, f, 1), (c, e, 8), (c, f, 10), (d, f, 1), (e, d, 4), (e, f, 7)\}$, where the notation $(a, b, 9)$ means that the edge from node a to node b has the weight 9.

In this graph, the goal is to find the longest path between nodes a and f. We adopt the solution heuristic based on selecting each time the longest edge among all edges originating from a given node. Starting from node a, we have two outgoing edges: $(a, b, 9)$ and $(a, c, 5)$, so we select edge $(a, b, 9)$.

Figure 1.8a. Using the same heuristic, in the following steps we choose edges $(b, d, 5)$ and $(d, f, 1)$. This situation is presented in Fig. 1.8b.

The path generated in this way consists of edges $((a, b, 9), (b, d, 5), (d, f, 1))$ and has length equal 15 ($9 + 5 + 1$), which can be shortly written as: $a \rightarrow b \rightarrow d \rightarrow f$ (15). It can be easily seen that this is not the best possible solution. In this example we can calculate the lengths (weights) of all possible paths from node a to node f as shown below:

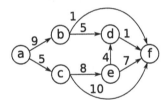

Fig. 1.7 Example graph G

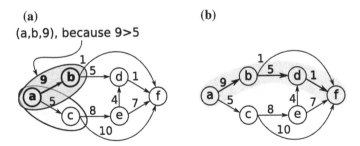

Fig. 1.8 Example path in graph G found using a greedy approach. **a** Selection of the first edge, **b** representation of whole path

$$a \to b \to f \qquad\qquad (10)$$
$$a \to b \to d \to f \qquad (15)$$
$$a \to c \to f \qquad\qquad (15)$$
$$a \to c \to e \to f \qquad (20)$$
$$a \to c \to e \to d \to f \ (18)$$

The approach based on stochastic algorithms allows us to find different solutions for the same problem without the need to check all possible paths. An example application of the ant colony optimization algorithm to the considered problem is presented in Fig. 1.9a. For simplicity, the heuristic function is omitted (we can be assume that its value equals 1 for every solution), and the initial pheromone value is set to 1.0.

In such a case, edge selection depends on the pheromone value, while edge weights are only used to evaluate the solution (although weight could also be used as the heuristic function). A higher pheromone value translates to a higher probability of selecting the given edge. Accordingly, the probability of selecting edges $(a, b, 9)$ and $(a, c, 5)$ in the first step is the same and equals 0.5.

Figure 1.9b presents an example of a single iteration of the ant colony optimization algorithm for a population consisting of 4 ant agents. We can see that in this solution at the beginning two ant agents would select edge $(a, b, 9)$, while edge $(a, c, 5)$ would be selected by the two remaining ant agents. In the next iteration of the algorithm, edge $(b, f, 1)$ would be selected by one agent, $(b, d, 1)$ would also be selected by one agent, and finally $(d, f, 1)$ would be selected as the only possibility. Similarly other ant agents would select edges $(c, f, 10)$ and $(c, e, 8)$, and afterwards $(d, f, 1)$ together with $(d, f, 1)$. It should be noted that in the first iteration the longest path was not set for any of the ant agents.

At the end of the example iteration, pheromone trail update is performed. To simplify: we can assume that every value of the pheromone trail is first evaporated by 10%, and then increased by the score value for each ant agent (for example, 10% of the sum of edge weights in the path which has been selected for the agent). In case of edge $(a, b, 9)$, the calculation is as follows:

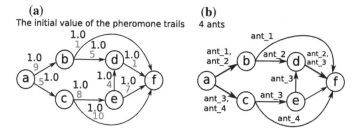

(a)
The initial value of the pheromone trails

(b)
4 ants

Fig. 1.9 Graph G in case of the ACO algorithm. **a** Initial values of the pheromone trail—edge representation. **b** Paths chosen for ant agents—edge thickness depends on the number of ant agents

$$1.0 - (1.0 \cdot 0.1) + \frac{10}{10} + \frac{15}{10} = 3.4,$$

As the initial value of the pheromone trail was equal 1.0, it is decreased by $(1.0 \cdot 0.1)$, and at the same time increased by the score (path) represented by the first $(\frac{10}{10})$ and the second $(\frac{15}{10})$ ant agents. The values of all edges after the pheromone trail update are given in Fig. 1.10a.

In the following steps of the ant colony optimization algorithm, such process is repeated many times. In this particular case, in the next iteration for edge $(a, c, 5)$ the selection probability should be increased (to 4.2 from 7.6; thus we can state that the edge would be selected in case of three ant agents). By proceeding similarly in the next stages of the algorithm, the longest path in the graph could be found.

Of course, the presented example is simplified—even in case of the small population of 4 ant agents, almost all possible solutions are covered. As we may guess, use of approximate algorithms is only reasonable in case of problems whose optimal solution is extremely difficult to find (for example, NP-hard problems).

In the presented example, another important issue can be seen. A lot of problems (including the learning process of decision trees) can be defined as a multiple decision problem in which at every stage a decision is taken—potentially the best one, but only at the specific stage. There is no certainty that any single decision is good in global terms.

The use of probability allows us to find solutions which can yield better results globally, even though other decisions were evaluated as better ones (at a single stage). Exactly such a situation is possible, for example, in the decision tree learning process based on the ant colony optimization algorithm. In case of using feedback (for example, pheromone) and repeating the decision processes, it is possible to store new, better single decisions, and at the same time adjust them for further stages of resolving the problem.

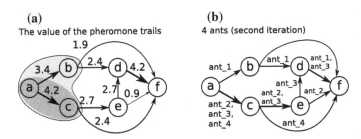

Fig. 1.10 The graph G in the second iteration of the ACO algorithm. **a** Pheromone trail values after the update, **b** paths selected for ant agents in the second iteration—thickness of the edge depends on the number of ant agents

1.3.3 Ant Colony Optimization Metaheuristic for Optimization Problems

The ACO metaheuristic is shown in Algorithm 1; it consists of an initialization step and a main loop over three algorithmic steps. A single iteration of the loop consists in all ants constructing solutions, their (optional) improvement with the use of a local search algorithm, and an update of the pheromones. In the following sections, we explain these three algorithmic components in more detail.

Algorithm 1: ACO metaheuristic for optimization problems

---

1 **procedure** Ant_colony_optimization_metaheuristic:
2   initialization();
3   **while** (termination_conditions_not_met)
4   construct_ant_solutions();
5   apply_local_search(); // optional
6   update_pheromones();
7   **endWhile**
8 **end** Ant colony optimization metaheuristic for optimization problems

---

initialization()

At the start of the algorithm, the parameters of this approach are established and all pheromone variables are initialized to a value τ_0, which is a parameter of this algorithm.

construct_ant_solutions()

A set of artificial ants construct solutions from elements of a finite set of available solution components $Comp$. Solution construction starts with an empty partial solution $sp = 0$. Then, at each construction step, the current partial solution sp is expanded by adding a feasible solution component from the set of feasible neighbors $N(sp) \subseteq Comp$. The process of formulating complete solutions can be regarded as a path in the construction graph or another part of a representation problem. The allowed or feasible solutions are implicitly defined by the solution construction mechanism that defines the set $N(sp)$ with respect to a partial solution sp.

The choice of a solution component from $N(sp)$ is made probabilistically at each construction step. The exact rules for the probabilistic choice of solution components vary across different ACO variants. The best known rule is that in the ant system (AS) [10, 13, 16], where τ is the pheromone value associated with a specific component, and η is a function that assigns a heuristic value to each feasible solution components belonging to $N(sp)$ at each construction step. The values that are returned by this function are commonly called the main probabilistic rules. Furthermore, α and β are positive parameters, whose values determine the relative importance of pheromone information versus heuristic information. In our version, the parameter α is equal to 1, and can therefore be omitted in (1.18), and subsequently, in (3.13).

The discussed procedure maintains the balance between the exploitation and exploration mechanisms for searching the solution space (i.e., the search space defined over a finite set of discrete decision variables, and the set of constraints among the variables). The complete construction of the solution takes into consideration both the feasibility and the differentiation between the individual solutions. In the next step, worthy solutions are not taken into consideration in the context of pheromone reinforcement.

The node transition rule is modified to explicitly allow for exploration. An ant k positioned at an analyzed node i chooses node j to move to, which can be written down using the following formula:

$$j = \begin{cases} \text{argmax}_{u \in J_i^k} \{[\tau_{iu}(t)] \cdot [\eta_{iu}]^\beta\} & \text{if } q \leq q_0 \\ J & \text{if } q > q_0 \end{cases}, \tag{1.18}$$

Here $\tau_{iu}(t)$ is the amount of pheromone currently available at time t (laid on the edge (i, u)), η_{iu} is the heuristic value between nodes i and u, q is a random variable uniformly distributed over $[0, 1]$, q_0 is a tunable parameter ($0 \leq q_0 \leq 1$), and $J \in J_i^k$ is the analyzed node, chosen randomly with the probability

$$p_{iJ}^k(t) = \frac{\tau_{iJ}(t) \cdot [\eta_{iJ}]^\beta}{\sum_{l \in J_i^k} [\tau_{il}(t)] \cdot [\eta_{il}]^\beta}, \tag{1.19}$$

which is similar to the transition probability used by the AS. Therefore, we can easily show that the ACS transition rule is identical to that of the AS for $q > q_0$, but different from it for $q \leq q_0$. More precisely, $q \leq q_0$ corresponds to the exploitation of available knowledge about the problem, i.e., heuristic knowledge about distances between the nodes, and the learned knowledge memorized in the form of pheromone trails, whereas $q > q_0$ favors more exploration.

apply_local_search()—optional
Once the solutions have been constructed, some additional actions may often be required before updating the pheromones. These are often called daemon actions, and can be used to implement problem-specific and/or centralized actions that cannot be performed by single ant agents. The most commonly used daemon action consists in applying local search to the constructed solutions. In other words, locally optimized solutions are used to decide which pheromone values should be updated in the pheromone matrix.

update_pheromones()
The aim of the pheromone update is to increase the pheromone values associated with good or promising solutions, and to decrease ("punish") those associated with bad solutions. Usually, this step is performed by decreasing all pheromone values through pheromone evaporation, and by increasing the pheromone levels associated with a chosen set of good solutions. These changes stimulate the process of reinforcement learning as observed in ACO.

In the AS, all ants are allowed to deposit the pheromone after completing their solutions. By contrast, in the ACO only the ant that has generated the best solution since the beginning of the trail is allowed to globally update the concentrations of pheromones on the branches. The updating rule is:

$$\Delta_{ij}(t+n) = (1-\gamma) \cdot \tau_{ij}(t) + \gamma \cdot \Delta\tau_{ij}(t, t+n), \qquad (1.20)$$

where (i, j) is the edge belonging to the best solution, γ is the parameter governing pheromone decay, and:

$$\Delta\tau_{ij}(t, t+n) = \frac{1}{ev}, \qquad (1.21)$$

where ev is the value of the evaluation of the best solution.

Local update is performed as follows: when, while constructing a solution, ant k is at an analyzed node i and selects node $j \in J_i^k$ to move to, the pheromone concentration on edge (i, j) is updated using the following formula:

$$\tau_{ij}(t+1) = (1-\rho) \cdot \tau_{ij}(t) + \rho \cdot \tau_0. \qquad (1.22)$$

The value of τ_0 is the same as the initial value of the pheromone trails. As described by [13, 17] it has been experimentally found that good results are obtained after setting $\tau_0 = (n \cdot U_{nn})^{-1}$, where n is the number of nodes and U_{nn} is the evaluation of the solution produced by the nearest neighbor's heuristic.

Pheromone evaporation is needed to avoid too rapid convergence of the algorithm. This method implements a useful form of forgetting, thus favoring the exploration of new areas in the search space.

1.4 Conclusions

In case of an introduction to a book like the present one, a balance between the two different disciplines discussed in it should be preserved. In both cases, a general introduction and more detailed solutions should be presented equally. The most important point of this book is the application of ant colony optimization to the problem of learning decision trees and ensemble methods. In this chapter we have presented a general introduction to machine learning concepts, as well as to swarm intelligence. Example solutions with some justification for the use of approximate methods in the decision tree learning process have been given as well. Ensemble learning will be discussed in detail in Chap. 6.

In addition to this chapter, the reader is referred to works on machine learning, like [27, 28, 32, 51], as a valuable source of information. The same applies to more detailed works on decision trees (including especially the CART algorithm, very important for this book) [37, 39, 56]. A range of works on the basics of swarm intelligence can be found in [3, 34]. The newest works on that subject and similar concepts are listed in [2, 60].

References

1. T. Bäck, D. Fogel, Z. Michalewicz, *Handbook of Evolutionary Computation* (Oxford, New York, 1997)
2. C. Blum, D. Merkle, *Swarm Intelligence: Introduction and Applications*, 1st edn. (Springer Publishing Company, Incorporated, 2008)
3. E. Bonabeau, M. Dorigo, G. Theraulaz, *Swarm Intelligence, From Natural to Artificial Systems* (Oxford University Press, 1999)
4. U. Boryczka, J. Kozak, Ant colony decision trees—a new method for constructing decision trees based on ant colony optimization, in *Computational Collective Intelligence. Technologies and Applications*, ed. by J.-S. Pan, S.-M. Chen, N. Nguyen. Lecture Notes in Computer Science vol. 6421 (Springer, Berlin/Heidelberg, 2010), pp. 373–382
5. L. Breiman, J.H. Friedman, R.A. Olshen, C.J. Stone, *Classification and Regression Trees* (Chapman & Hall, New York, 1984)
6. R.R. Bulatović, G. Miodragović, M.S. Bošković, Modified krill herd (mkh) algorithm and its application in dimensional synthesis of a four-bar linkage. Mech. Mach. Theory **95**, 1–21 (2016)
7. D. Corne, M. Dorigo, F. Glover, *New Ideas in Optimization* (Mc Graw-Hill, Cambridge, 1999)
8. J.L. Deneubourg, S. Goss, N.R. Franks, J.M. Pasteels, The blind leading the blind: modelling chemically mediated army ant raid patterns. Insect Behav. **2**, 719–725 (1989)
9. K.F. Doerner, D. Merkle, T. Stützle, Special issue on ant colony optimization. Swarm Intell. **3**(1), 1–2 (2009)
10. M. Dorigo, Optimization, learning and natural algorithms (in Italian). Ph.D. thesis, Dipartimento di Elettronica, Politecnico di Milano, IT, 1992
11. M. Dorigo, M. Birattari, T. Stützle, Ant colony optimization—artificial ants as a computational intelligence technique. IEEE Comput. Intell. Mag. **1**, 28–39 (2006)
12. M. Dorigo, G. Di Caro, L.M. Gambardella, Ant algorithms for distributed discrete optimization. Artif. Life **5**(2), 137–172 (1999)
13. M. Dorigo, L.M. Gambardella, A study of some properties of Ant–Q, in *Proceedings of Fourth International Conference on Parallel Problem Solving from Nature, PPSNIV* (Springer, Berlin, 1996), pp. 656–665
14. M. Dorigo, L.M. Gambardella, Ant colony system: a cooperative learning approach to the traveling salesman problem. IEEE Trans. Evol. Comp. **1**, 53–66 (1997)
15. M. Dorigo, V. Maniezzo, A. Colorni, Positive feedback as a search strategy. Technical Report 91–016, Politechnico di Milano, Italy, 1991
16. M. Dorigo, V. Maniezzo, A. Colorni, The ant system: an autocatalytic optimization process. Technical Report 91-016, Department of Electronics, Politecnico di Milano, Italy, 1996
17. M. Dorigo, T. Stützle, *Ant Colony Optimization* (MIT Press, Cambridge, 2004)
18. R. Eberhart, J. Kennedy, A new optimizer using particle swarm theory, in *Proceedings of the Sixth International Symposium on Micro Machine and Human Science, 1995. MHS'95* (IEEE, 1995), pp. 39–43
19. F. Esposito, D. Malerba, G. Semeraro, A comparative analysis of methods for pruning decision trees. IEEE Trans. Pattern Anal. Mach. Intell. **19**, 476–491 (1997)
20. I. Fister Jr, X. Yang, I. Fister, J. Brest, D. Fister, A brief review of nature-inspired algorithms for optimization (2013). arXiv:1307.4186
21. L.J. Fogel, A.J. Owens, M.J. Walsh, *Artificial Intelligence through Simulated Evolution* (Wiley, New York, 1966)
22. L. M. Gambardella, M. Dorigo, HAS–SOP: Hybrid Ant System for the Sequential Ordering Problem. Technical Report 11, IDSIA, 1997
23. A.H. Gandomi, A.H. Alavi, Krill herd: a new bio-inspired optimization algorithm. Commun. Nonlinear Sci. Numer. Simul. **17**(12), 4831–4845 (2012)
24. F. Glover, M. Laguna, *Tabu Search* (Kluwer Academic Publishers, Dordrecht, 1997)
25. P.P. Grasse, La reconstruction du nid et les coordinations inter–individuelles chez bellicositermes natalensis et cubitermes sp. La theorie de la stigmerie. Insects Soc. **6**, 41–80 (1959)

26. P.P. Grasse, *Termitologia*, vol. II (Masson, Paris, 1984)
27. M. Hall, I. Witten, E. Frank, *Data Mining: Practical Machine Learning Tools and Techniques* (Kaufmann, Burlington, 2011)
28. J. Han, J. Pei, M. Kamber, *Data Mining: Concepts and Techniques* (Elsevier, 2011)
29. T. Hastie, R. Tibshirani, J. Friedman, *The Elements of Statistical Learning*. Springer Series in Statistics (Springer New York Inc., New York, NY, USA, 2001)
30. J.H. Holland, *Adaptation in Natural and Artificial Systems* (University of Michigan Press, Ann Arbor, 1975)
31. L. Hyafil, R.L. Rivest, Constructing optimal binary decision trees is np-complete. Inf. Process. Lett. **5**(1), 15–17 (1976)
32. M. Kantardzic, *Data Mining: Concepts, Models, Methods, and Algorithms* (Wiley, 2011)
33. D. Karaboga, An idea based on honey bee swarm for numerical optimization. Technical report, Technical report-tr06, Erciyes university, engineering faculty, computer engineering department, 2005
34. J.F. Kennedy, J. Kennedy, R.C. Eberhart, Y. Shi, *Swarm Intelligence* (Morgan Kaufmann, 2001)
35. J. Kiefer, J. Wolfowitz, Stochastic estimation of the maximum of a regression function. Ann. Math. Stat. **23**(3), 462–466 (1952)
36. S. Kirkpatrick, C.D. Gelatt, M.P. Vecchi, Optimization by simulated annealing. Science **220**(4598), 671–680 (1983)
37. S.B. Kotsiantis, Decision trees: a recent overview. Artif. Intell. Rev. **39**(4), 261–283 (2013)
38. D.T. Larose, *Discovering Knowledge in Data: An Introduction to Data Mining* (Wiley-Interscience, 2004)
39. W. Loh, Classification and regression trees. Wiley Interdisc. Rev.: Data Min. Knowl. Disc. **1**(1), 14–23 (2011)
40. Z. Michalewicz, D. Fogel, *How to Solve It: Modern Heuristics* (Springer, Berlin, Heidelberg, 2004)
41. J. Mingers, An empirical comparison of pruning methods for decision tree induction. Mach. Learn. **4**(2), 227–243 (1989)
42. T.M. Mitchell, Machine learning. wcb, 1997
43. P. Moscato, On evolution, search, optimization, genetic algorithms and martial arts: towards memetic algorithms. Caltech concurrent computation program, C3P. Report 826:1989, 1989
44. O.J. Murphy, R.L. McCraw, Designing storage efficient decision trees. IEEE Trans. Comput. **40**(3), 315–320 (1991)
45. T. Niblett, I. Bratko, Learning decision rules in noisy domains, in *Proceedings of Expert Systems '86, The 6Th Annual Technical Conference on Research and development in Expert Systems III* (New York, NY, USA, 1987), pp. 25–34
46. I. Osman, G. Laporte, Metaheuristics: a bibliography. Ann. Oper. Res. **63**, 513–623 (1996)
47. R.S. Parpinelli, H.S. Lopes, A.A. Freitas, An ant colony algorithm for classification rule discovery, in *Data Mining: A Heuristic Approach*, ed. by H. Abbas, R. Sarker, C. Newton (Idea Group Publishing, London, 2002), pp. 191–208
48. J.R. Quinlan, Induction of decision trees. Mach. Learn. **1**(1), 81–106 (1986)
49. J.R. Quinlan, Simplifying decision trees. Int. J. Man-Mach. Stud. **27**(3), 221–234 (1987)
50. J.R. Quinlan, *C4.5: Programs for Machine Learning* (Morgan Kaufmann, 1993)
51. B. Ratner, *Statistical and Machine-Learning Data Mining: Techniques for Better Predictive Modeling and Analysis of Big Data* (CRC Press, 2011)
52. C. Reeves, Modern heuristic techniques for combinatorial problems, *Advanced Topics in Computer Science* (McGrawHill, London, 1995)
53. H. Robbins, S. Monro, A stochastic approximation method. Ann. Math. Stat. 400–407 (1951)
54. L. Rokach, O. Maimon, *Data Mining With Decision Trees: Theory and Applications* (World Scientific Publishing, 2008)
55. B. Rylander, T. Soule, J. Foster, J. Alves-Foss, Quantum genetic algorithms, in *Proceedings of the 2nd Annual Conference on Genetic and Evolutionary Computation* (Morgan Kaufmann Publishers Inc., 2000), pp. 373–373

56. R. Timofeev, Classification and regression trees (CART) theory and applications. Master's thesis, CASE Humboldt University, Berlin, 2004
57. G. Wang, S. Deb, S.M. Thampi, A discrete krill herd method with multilayer coding strategy for flexible job-shop scheduling problem. Intell. Syst. Technol. Appl. 201–215 (2016)
58. G. Wang, A.H. Gandomi, A.H. Alavi, S. Deb, A hybrid method based on krill herd and quantum-behaved particle swarm optimization. Neural Comput. Appl. **27**(4), 989–1006 (2016)
59. Y. Yang, G. Scutari, D.P. Palomar, M. Pesavento, A parallel stochastic approximation method for nonconvex multi-agent optimization problems (2014). arXiv:1410.5076
60. Z. Zhang, K. Long, J. Wang, F. Dressler, On swarm intelligence inspired self-organized networking: its bionic mechanisms, designing principles and optimization approaches. IEEE Commun. Surv. Tutor. **16**(1), 513–537 (2014)

Part I
Adaptation of Ant Colony Optimization to Decision Trees

Chapter 2
Evolutionary Computing Techniques in Data Mining

2.1 Introduction

A single method cannot be the best approach to solving every problem. One way to deal with this issue is to hybridize an algorithm with more standard procedures, such as greedy methods or local search procedures. The individual solutions obtained so far can be improved by using other search procedures, and then put back in competition with other solutions that have not been improved yet.

There are so many ways to hybridize algorithms that there is a common tendency to overload hybrid systems with too many components. Some of the modern hybrid systems contain fuzzy-neural-evolutionary components, together with other problem-specific heuristics. Ant colony optimization algorithms are extremely flexible, and can be extended by incorporating diverse concepts and alternative approaches. For example, Stüzle and Hoos [93] have implemented a version of AS-QAP based on their MAX-MIN Ant System (MMAS). They discussed the possibility of incorporating a local search (2-opt algorithm and short Tabu Search runs) into the ant algorithm in case of QAP. The ant colony optimization technique can also be enhanced by inserting new steps into the standard version of this algorithm.

However, in this chapter we present mostly solutions that combine swarm intelligence algorithms (not only ACO) with the classical data mining algorithms. Particular emphasis is placed on algorithms for decision rules, clustering and decision trees—each time with an attempt to present ant colony optimization algorithms. We should mention here that similar analysis concerning ensemble methods is presented in Sect. 6.5.

© Springer International Publishing AG, part of Springer Nature 2019
J. Kozak, *Decision Tree and Ensemble Learning Based on Ant
Colony Optimization*, Studies in Computational Intelligence 781,
https://doi.org/10.1007/978-3-319-93752-6_2

2.2   Decision Rules

In the literature there are several articles discussing ant colony optimization algorithms that concern decision rules. One of the most popular methods of this kind is the Ant-Miner algorithm. Such a combination with ant colony optimization (in principle, the ant colony algorithm) was proposed in 2002 by Parpinelli et al. [71]. The results obtained by the authors as compared to CN2 [28] and C4.5 [73] indicated an improvement in selected results, and more of them were published in further articles [71, 72]. The authors' work was continued, and further analysis and propositions can be found, for example, in [36, 58, 62, 91].

Despite its use in the heuristics of solutions used for building decision trees (C4.5), Ant-Miner is also an algorithm for induction of decision rules. This enforces certain specific solutions related to a different interpretation of the heuristic function quality. Under that interpretation, in contrast to decision trees, lower entropy of a descriptor in the data sets gives a higher probability of its better quality. In addition to that issue, another specific element of the algorithm is the way of updating the pheromone trail. Each time after the pheromone is laid, normalization of all its values occurs. Although this causes its decrease (with respect to the heuristic function value), the importance of the time at which the pheromone values were laid is not marked in any way. This indicates incomplete use of the potential of ant colony algorithms. Moreover, Ant-Miner in its classical version is an ant algorithm, so its population is a single ant agent, and the algorithm is not suitable for data with continuous attributes.

The authors of Ant-Miner [71, 72] suggested two directions for future research: extension of Ant-Miner to cope with continuous attributes and investigation of the effects of changes in the main transition rule. This chapter gives also a brief overview of Ant-Miner extensions. Some of the proposals are relatively simple and, as a result, give the same type of classification rules discovered by this algorithm. Other modifications cope with the problem of attributes with ordered categorical values; some of them improve the flexibility of the rule representation language. Finally, more sophisticated modifications have been proposed to discover multi-label classification rules, and to investigate fuzzy classification rules. Certainly, there are still many problems and open questions for future research. Modifications and extensions to Ant-Miner have been presented in many articles, as listed in Table 2.1.

Many of the algorithms described in Table 2.1 as modifications and extensions of Ant-Miner can be qualified as separate ant colony algorithms for building decision rules. These are mostly algorithms proposed by Liu et al. (Ant-Miner 3) [58], Wang and Feng (ACO-Miner) [99], Jiang et al. (Ant-Miner(I)) [45], Martens et al. (Ant-Miner+) [62, 63], or Chan and Freitas (MuLAM) [21], as well as various versions of an algorithm simulating the parallel algorithm proposed by Chen and Tu [25], and also by Roozmand and Zamanifa [25, 75].

Moreover, other ant colony optimization algorithms for building classifiers that arise from Ant-Miner should be emphasized. However, the changes introduced in that approach are significant. Among the above algorithms, there are those proposed by Holden and Freitas [40, 41] PSO/ACO and PSO/ACO2. By creating a hybrid

Table 2.1 Overview of the articles on Ant-Miner

1. Modifications or extensions: extension of the Ant-Miner is one of the crucial issues in its adaptation to data with continuous attributes [42, 68, 69, 80, 86]

Authors	Year	Conclusions and open questions
Otero, Freitas & Johnson, Schaefer	2008, 2009, 2013, 2009	The proposed modification was a new version of the algorithm ($cAnt-Miner$, next $cAnt-Miner_{PB}$), where discretization of continuous attributes based on entropy was introduced. This approach was later extended, and additional discretization methods were proposed. Experiments show that discretization carried out by the algorithm improves the results obtained

2. Modifications or extensions: new method of pheromone trail update. Changes in pheromone evaporation speed, as well as in available pheromone trail update methods [45, 58, 62, 63, 91, 99]

Authors	Year	Conclusions and open questions
Jiang, Xu & Xu, Liu, Abbas & Mc Kay, Martens et al., Smalton & Freitas, Wang & Feng, Salama et al., Hota et al.	2005, 2004, 2006, 2007, 2006, 2004, 2013, 2014	Change in the pheromone trail was important due to the results obtained with the algorithm. A new method of pheromone trail update, giving a greater importance to the pheromone evaporation concept, should be proposed the method for updating the pheromone trail in such way that the pheromone value is stored on a current basis during the process of rule building by every ant agent

3. Modifications or extensions: introduction of rule quality score. Replacement of the sensitivity and specificity measures employed in the employed previously in rule quality evaluation with new measures: confidence and coverage [22, 62]

Authors	Year	Conclusions and open questions
Chen, Chen & He, Martens et al.	2006, 2006	To confirm the rationale for such a quality measurement, a more extensive group of experiments should be carried out in order to enable comparison of the results obtained using the new and old approaches

4. Modifications or extensions: new heuristic function. Replacement of the complex heuristic function based on entropy with new, simpler methods for calculating the heuristic function [6, 22, 45, 58, 59, 62, 63, 67, 91, 99]

Authors	Year	Conclusions and open questions
Chen, Chen & He, Jiang, Xu & Xu, Liu, Abbas & Mc Kay, Martens et al., Oakes, Wang & Feng, Boryczka & Kozak	2006, 2005, 2006, 2007, 2004, 2006, 2004, 2009	Application of the pheromone trail alone (without a heuristic function) does not allow for achieving results similar to those of the standard algorithm. Single-time calculation of the heuristic function—only during algorithm initialization—may be proposed. Moreover, it can be made dependent on the data set structure (number of objects—examples, number of attributes). Furthermore, Boryczka & Kozak proposed application of a heuristic function based on different algorithms. The proposed example solutions known from the CART and CN2 algorithms—especially those achieved using the heuristic function from the CART algorithm—yielded very good results

(continued)

Table 2.1 (continued)

5. Modifications or extensions: pseudorandom proportional rule (main transition rule) using the exploration/exploitation division (parameter q_0) [22, 45, 58, 99]

Authors	Year	Conclusions and open questions
Chen, Chen & He, Jiang, Xu & Xu, Liu, Abbas & Mc Kay, Wang & Feng	2006, 2005, 2004, 2004	Application of clear exploration/exploitation division gives a lot of possibilities. However, a big problem is still selection of the appropriate values of parameter q_0

6. Modifications or extensions: discovery of fuzzy rules by employing multiple populations of ant agents, each of which discovers a rule belonging to a different class (but compatible with class for the whole colony) or by proposing a version of the algorithm with a different heuristic [36, 61]

Authors	Year	Conclusions and open questions
Galea & Shen, Madhusudhanan, Karnan & Rajivgandhi	2006, 2010	The created set of rules is smaller than that acquired during the operation of the classical algorithm—only the potentially best rules are included. The rules are more readable for the human, and can be used as support for the human factor. However, the decisions should be always verified

7. Modifications or extensions: a decision class is set for the created rule, and all ant agents construct rules for the same decision class [22, 36, 62, 63, 91]

Authors	Year	Conclusions and open questions
Chen, Chen & He, Galea & Shen, Martens et al., Smaldon & Freitas	2006, 2006, 2007, 2006	Such an approach leads to the necessity of creating a new, different heuristic function and a new approach to pheromone trail update

8. Modifications or extensions: simultaneous modification of multiple elements of the algorithm to improve its efficiency [7, 76]

Authors	Year	Conclusions and open questions
Salama & Abdelbar, Boryczka & Kozak	2010, 2009	Introduction of conjunction into the premise of a rule, a specific ant agent along with various modifications in the pheromone trail, as well as introduction of individual parameter values for ant agents allows for improving the quality of the generated rules. The authors proved empirically that the developed rules are shorter than the classical ones while maintaining good classification results. In the work Boryczka & Kozak, many precise analyses of Ant-Miner can be found. They include application of the pheromone trail, its initial values, heuristic function and so on. It is a one of the most important papers on the analysis of Ant-Miner

(continued)

Table 2.1 (continued)

9. Modifications or extensions: a parallel Ant-Miner algorithm (simulation) [13, 25, 75]

Authors	Year	Conclusions and open questions
Chen, Tu, Roozmand & Zamanifar, Boryczka & Kozak	2005, 2008, 2009	The Ant-Miner algorithm is still used as a sequential one, but a simulation of the parallel concept was used. A parallel version is based on executing calculations for groups of ant agents, where each group builds rules for a given decision class. After a predefined time, communication between the groups is carried out based on the update (combination) of the pheromone trails of all groups. The authors indicated the legitimacy of such an approach, especially in the case of acceleration. However, for more in-depth experiments, a fully parallel version of Ant-Miner is needed

10. Modifications or extensions: multi-class learning, in which an object being classified can be assigned to more than one decision class [21]

Authors	Year	Conclusions and open questions
Chan & Freitas	2006	Application of a pheromone matrix for every decision class was proposed. During pheromone trail update, various dependencies among the decision classes are analyzed

11. Modifications or extensions: the algorithm is adapted to ranging over discrete attribute values (rather than continuous ones) [62, 67]

Authors	Year	Conclusions and open questions
Oakes, Martens et al.	2004, 2006	The important issue is showing that combination with rough set theory, in contrast to combination with fuzzy sets, is a promising approach

12. Modifications or extensions: changes related to trimming (shortening) rules [62]

Authors	Year	Conclusions and open questions
Martens et al.	2006	Algorithm operating without rule trimming is significantly faster, but results in generating very long, useless rules. However, trimming of the rules can be applied after the operation of Ant-Miner is completed

algorithm based on a combination of PSO and ACO, the authors enable, for example, working with continuous values and nominal values without the necessity to convert them into numerical values. An algorithm worth noting is TACO-Miner proposed by Thangaveli and Jaganathan [96]. By introducing general changes, its authors tried to achieve better classification accuracy and to simplify the constructed set of rules. Following this, in 2011 Salama et al. developed Multi-pheromone Ant-Miner [79, 80], in which the idea of a class-based pheromone was proposed. In this case, an ant agent is assigned to a decision class, and during the whole algorithm run the rules are built for particular classes (every class is based on a different pheromone trail matrix). Such juxtaposition was extended for example by Chircop and Buckingham, who proposed Multiple Pheromone Ant Clustering Algorithm [27].

Furthermore, there are also publications on applications of Ant-Miner to particular problems. A good example of such an application is the algorithm used for website classification [39] and broadly understood medicine and bio-medicine—for example, the currently popular analysis of gene expression and disease diagnosis [29, 56, 74, 87, 98].

In the presented juxtaposition, a few very important genetic and evolutionary algorithms should also be mentioned. Those approaches were already popular in the 1990s—see e.g. [31, 33, 57, 60]. However, the discussed subject is so broad that, despite the numerous genetic and evolutionary algorithms for rule induction, those mentioned above represent only a small fraction of all works on that topic. A good study in this range is the book [34]. Other collective intelligence algorithms used in rule induction are artificial bee colony—and, more specifically, Bee-Miner [19] proposed in 2011. An algorithm of the same name was also proposed in 2016 [95]. However, it was based on a different group of algorithms—so-called bees algorithms.

2.3 Decision Trees

As to learning decision trees, only the ACDT algorithm (the main subject of this book)—first proposed in 2010—and Ant-Tree-Miner—presented in 2012—were introduced as ACO based learning. So far, the use of ACO in decision tree construction was first described in [8]. These were preliminary versions of ACDT (its current version is described in Chap. 3).

ACDT was later extended, and up to now has been presented by Kozak et al. in several articles. Soon after the first publication, a new ACDT version adapted to data sets with continuous parameters was presented in [9] (its detailed description can be found in Chap. 3). Next, also in [10], the parameters used in ACDT were analyzed, with particular emphasis on q_0. The mentioned parameter governs the exploration/exploitation mechanism in ACO. That research allowed for estimating the values of some parameters. In turn, [14] presented a parallel ACDT algorithm for constructing decision trees, in which several ant colonies cooperated periodically. Ants in the colonies cooperated on two levels: intra-colony—using a pheromone trail, and inter-colony—using explicit information exchange in the form of the best quality solutions found. Two inter-colony cooperation strategies were proposed and tested for classification accuracy and computation performance.

The newest version of ACDT, described in Chap. 3, was published in 2016. In [49], the authors presented a formal definition of the algorithm, and described all solutions in detail. The aim of [49] was to determine the role played in ACDT by the pheromone relative to the heuristic function. Moreover, the authors wanted to analyze the cooperation and communication between the ant agents via the pheromone trail. In turn, the main aspect of [11] was to examine various ways of evaluating the size of decision trees, and to find a compromise between measuring the classification quality and the decision tree size. Part of those observations are included in Chap. 3). A separate chapter is devoted to a special version of ACDT, where the goal function can be changed depending on the target of the constructed classifier. Detailed research on the adaptive goal function of ACDT is included in [48], and described in Chap. 4.

After certain modifications, ACDT can be effectively applied to various practical problems, such as estimation of dry docking duration [94], automatic categorization of email [15–17], and discovery of financial data rules [50]. ACDT has also been applied in H-bond data set analysis [47]. Some of the ACDT applications will be presented in detail in Chap. 5.

Otero et al. [70] proposed a different algorithm based on ACO for decision tree induction. The idea is similar (pseudo-code is the same as for Algorithm 2), although the authors of Ant-Tree-Miner applied other techniques. Typically, the information gain ratio from the C4.5 approach was used as the heuristic function. That approach does not construct binary decision trees. To generalize the classifier, authors applied two-step pruning of the prepared decision trees, which was somewhat similar to the method applied in the C4.5 algorithm. Decision tree quality based on pheromone trail updating corresponds to the classification accuracy of the training set objects. An important aspect of Ant-Tree-Miner is also the pheromone matrix, which stores in the nodes information not only about attributes and their values, but also about the decision tree level where the node is located. That structure is one significant difference between the Ant-Tree-Miner and ACDT approaches.

Next, Otero et al. proposed learning multi-tree classification based on ant colony optimization [85]. In this case, for every decision class a single decision tree is created. Thus the multi-tree model consists of different decision trees $DT_1, DT_2, \ldots, DT_C$, where C is the number of decision classes. In addition, DT_l is responsible for distinguishing class l from all the remaining decision classes. As a result, the authors proposed two algorithms based on Ant-Tree-Miner: a local ACO approach for learning multi-trees and an integrated ACO approach for learning multi-trees. The experiments they conducted showed an improvement in the results over Ant-Tree-Miner.

Research focused on learning decision trees with the use of ACO algorithms seems to be almost unknown: there are other methods devoted to application of computational intelligence in decision tree induction. More solutions, including the use of genetic algorithms and evolutionary algorithms in decision tree construction, can be found in the literature. We should note that, in contrast to ACO, evolutionary algorithms do not use local heuristics, and their search is conducted only on the basis of a fitness function (for example, global quality of the decision tree). In the ACO algorithms, search is conducted on the basis of general solution quality, and according to local heuristic information. Additionally, a kind of feedback is the pheromone trail level, which allows for local search near potentially good solutions.

Evolutionary techniques are most often used to construct decision trees. The first approaches to constructing binary decision trees were almost simultaneously presented by Chai et al. [20] and Kennedy et al. [46], but work on this subject dates back to the early 1990s. Subsequent studies using genetic algorithms were carried out by Fu et al. [35], who proposed a method where the base population consisted of trees that had been constructed using the C4.5 algorithm, and on which genetic operations had been performed. However, at the last stage of the algorithm, the trees were corrected and pruned.

The memetic algorithm for decision tree induction proposed by Krętowski [51, 52] was an extension of the previously proposed evolutionary algorithm for constructing decision trees. The global decision tree axis-parallel (GDT-AP) is an evolutionary algorithm used in global induction of decision trees (single-dimension). It is the basic version of an algorithm widely described, and then extended in [51, 53, 54]. In turn, the global decision tree memetic algorithm (GDT-MA) is a hybrid algorithm created by combining an evolutionary algorithm with a local search technique used in the induction of decision trees, described based on the GDT-AP algorithm. Compared to GDT-AP, the method of population creation was changed, along with the mutation operator. A detailed description of the algorithm can be found in [52]. Recently, Jankowski and Jackowski proposed the EVO-tree algorithm [44]. EVO-Tree uses an evolutionary algorithm for induction of decision trees as a tool for minimizing a global function based on the decision tree size and accuracy.

A very good work on evolutionary algorithms for decision tree induction was written by Barros et al. [5]. The authors analyzed over 100 articles on that subject, and briefly characterized selected articles. This review work could be a good extension of the present chapter.

For the purpose of constructing smaller decision trees than those generated by genetic algorithms, algorithms based on genetic programming were also proposed. Thus, Niimi et al. [65] presented a hybrid approach based on a combination of genetic programming and association rules. In turn, [32] proposed a combination of genetic programming with simulated annealing, which was used by Heath et al. [38] to construct decision trees. The authors of those articles also used the C4.5 algorithm as a basis, and compared the obtained results with those yielded by that algorithm.

It is difficult to find other swarm intelligence algorithms used for learning decision trees. There are a few solutions where swarm intelligence algorithms are used for decision trees, but they are not closely related to the process of constructing those trees. They include, for example, the algorithms with earlier data fitting, frequently mentioned in the literature, which are used as a basis for constructing decision trees [2, 24, 100]. An artificial ant colony system for building a regression tree, proposed and described in 2001 by Izaraiev and Agrafiotis in [43], can be mentioned as well. A regression tree may be treated as a special case of a decision tree, where the leaves (decision classes) have continuous values. In the proposed method, the authors introduced the use of pheromone in the process of selecting the attribute and its values at every single decision node. However, the described algorithm had two major disadvantages: it could only be used for attributes with continuous values (of course, the decision attribute should be continuous as well), and it did not use the heuristic function, which is extremely important in ant colony optimization algorithms.

2.4 Other Algorithms

Besides the concepts discussed in this chapter and those presented in Table 2.1, ant colony optimization algorithms are used in other knowledge data discovery (KDD) problems. The literature contains numerous proposals for such use. The examples include algorithms for learning various types of Bayesian network classifiers, like ABC-Miner proposed in 2013 by Salama and Freitas [83]. The first mentioned version of ABC-Miner learns the Bayesian Augmented Naive-Bayes (BAN) classifier. The same authors in [84] proposed ABC-Miner+, which is used in the process of creating Markov Blanket classifiers. In turn, ACO for learning Bayesian multi-nets was proposed in [81, 82].

Ant colony optimization algorithms are widely used in the clustering process. Undoubtedly, ant based clustering is one of the most popular concepts among swarm based clustering algorithms. An algorithm for optimal clustering of N objects into K clusters was proposed already in 2004. Shelokar et al. tested it in [89], confirming that the considered approach gave positive results. More information on the use of ACO in clustering can be found in [37], where Handl et al. conducted a thorough analysis of ant based clustering algorithms. They divided them into two categories: the first included methods directly imitating real ant colonies, and the second—algorithms more significantly based on heuristic solutions of the classical algorithms. Among all publications, an article worth mentioning is [27], which described the MPACA algorithm. In that algorithm, the main novelty was the use of multi-pheromone trails (one for every value of the attribute). In that solution, every object, or in fact every attribute, was assigned some ant agents. The algorithm used them to find other objects with the same values of the attribute. ACO based clustering is still being developed, and one of the newest papers can be found in the [12, 77].

It is not difficult to find proposals of ant algorithms for the data preparation process—and, more specifically, instance selection/data reduction. One of those proposals is the ADR-Miner algorithm introduced by Anwar et al., and later extended and described in [3]. An alternative algorithm for data reduction based on ACO was also proposed by Salama et al. in [78].

Other, less popular swarm intelligence algorithms are used for data mining. An example of such methods is the grey wolf optimizer in a hybrid combination with random forests [92], or the cat swarm optimization in the clustering process (here with the more known PSO) [55]. The mentioned PSO has been used in clustering since the early 21st century [97] (where two PSO algorithms connected with k-means were proposed), and is still being developed [23]. In the literature we can find examples of a combination of ACO and PSO used in the clustering process. Such a case is e.g.

[66], where a hybrid of three algorithms was proposed: PSO (more exactly, fuzzy adaptive particle swarm optimization), together with ACO and k-means. A detailed review of PSO algorithms on that subject can be found in [30].

Of course, similarly like in case of Sect. 2.2, genetic algorithms and evolutionary algorithms are also represented by multiple solutions for the considered problems, discussed in numerous papers (beginning with [90]). Genetic algorithms have often been used in clustering along with k-means [4, 26, 64, 88].

2.5 Conclusions

The review of works presented in this chapter contains only the core algorithms related to evolutionary computing techniques in data mining, and their examples. According to the subject of this book, articles on ACO based learning are the crucial ones, and the remaining methods concerning swarm intelligence represent a kind of background only. There is no way to present here an analysis of the whole domain of knowledge data discovery. Thus the greatest emphasis has been put on the articles that are important from the viewpoint of this book—concerning decision trees and decision rules, which in case of ant colony optimization algorithms are strongly connected with the subject of learning decision trees.

To present the large collection of papers on the discussed subject, Fig. 2.1 shows a review of data mining algorithms based on swarm intelligence (by the number of algorithms/number of publications on that subject). For any technique, the estimated number of publications on its subject is presented. As the numbers represent a kind of order of magnitude, they allow us to see some trends. However, the exact number of publications on a given subject is difficult to estimate. The figure presents certain milestones as well—they include algorithms or approaches important from the view-

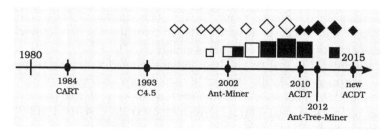

Ant-Miner - ACO (decision rules)
ACDT - ACO (decision trees)
Ant-Tree-Miner - ACO (decision trees)
  ☐ - Decision rules
  ◇ - Decision trees
  ■◆ - Ant Colony Optimization

Fig. 2.1 Data mining algorithms based on swarm intelligence

point of this book. In this case, by way of exception they also include the classical ones, like C4.5 [73] and CART [18].

We should also mention here the book [1], whose editors (Abraham et al.) tried to present a review of swarm intelligence in data mining, with a particular emphasis on papers discussing ant colony optimization algorithms. That book can be treated as a kind of supplement to the present one.

Moreover, to observe the adopted structure of the book—the review of articles on ensemble learning has mostly been omitted here, and can be found in the second part of the book, directly concerned with ACO based ensemble learning.

References

1. A. Abraham, C. Grosan, V. Ramos. *Swarm Intelligence in Data Mining*, vol. 34 (Springer, 2007)
2. T. Amin, I. Chikalov, M. Moshkov, B. Zielosko, Dynamic programming approach to optimization of approximate decision rules. Inf. Sci. **221**, 403–418 (2013)
3. I.M. Anwar, K.M. Salama, A.M. Abdelbar, Instance selection with ant colony optimization. Proc. Comput. Sci. **53**, 248–256 (2015)
4. S. Bandyopadhyay, U. Maulik, Genetic clustering for automatic evolution of clusters and application to image classification. Pattern Recogn. **35**(6), 1197–1208 (2002)
5. R.C. Barros, M.P. Basgalupp, A. De Carvalho, A.A. Freitas, A survey of evolutionary algorithms for decision-tree induction. IEEE Trans. Syst Man Cybern Part C (Applications and Reviews) **42**(3), 291–312 (2012)
6. U. Boryczka, J. Kozak, A new heuristic function in ant–miner approach, in *11th International Conference on Enterprise Information Systems—ICEIS 2009; ISBN: 978-989-8111-85-2* (Milan, Italy, 2009), pp. 33–38
7. U. Boryczka, J. Kozak, New algorithms for generation decision trees–ant-miner and its modifications, in *Foundations of Computational Intelligence*, vol. 6 (Springer, Berlin, Germany, 2009), pp. 229–264
8. U. Boryczka, J. Kozak, Ant colony decision trees–a new method for constructing decision trees based on ant colony optimization, in *Computational Collective Intelligence. Technologies and Applications*. Lecture Notes in Computer Science, ed. by J.-S. Pan, S.-M. Chen, N. Nguyen, vol. 6421 (Springer, Berlin/Heidelberg, 2010), pp. 373–382
9. U. Boryczka, J. Kozak, An adaptive discretization in the acdt algorithm for continuous attributes, in *Computational Collective Intelligence. Technologies and Applications. LNCS*, vol. 6923 (Springer, 2011), pp. 475–484
10. U. Boryczka, J. Kozak, New insights of cooperation among ants in ant colony decision trees, in *Third World Congress on Nature & Biologically Inspired Computing, NaBIC 2011, Salamanca, Spain, October 19–21, 2011* (2011), pp. 255–260
11. U. Boryczka, J. Kozak, Enhancing the effectiveness of ant colony decision tree algorithms by co-learning. Appl. Soft Comput. **30**, 166–178 (2015)
12. U. Boryczka, J. Kozak, Adaptive ant clustering algorithm with pheromone, in *Asian Conference on Intelligent Information and Database Systems* (Springer, Berlin, Heidelberg, 2016), pp. 117–126
13. U. Boryczka, J. Kozak, R. Skinderowicz, Parellel ant–miner. Parellel implementation of an ACO techniques to discover classification rules with OpenMP, in *15th International Conference on Soft Computing—MENDEL 2009; ISBN: 978-80-214-3884-2* (Brno, Czechy, Brno University of Technology 2009), pp. 197–205
14. U. Boryczka, J. Kozak, R. Skinderowicz, Heterarchy in constructing decision trees—parallel acdt. T. Comp. Collective Intell. **10**, 177–192 (2013)

15. U. Boryczka, B. Probierz, J. Kozak, An ant colony optimization algorithm for an automatic categorization of emails, in *Computational Collective Intelligence. Technologies and Applications—6th International Conference, ICCCI 2014, Seoul, Korea, September 24–26, 2014* (2014), pp. 583–592

16. U. Boryczka, B. Probierz, J. Kozak, A new algorithm to categorize e-mail messages to folders with social networks analysis, in *Computational Collective Intelligence* (Springer International Publishing, 2015), pp. 89–98

17. U. Boryczka, B. Probierz, J. Kozak, Automatic categorization of email into folders by ant colony decision tree and social networks, in *Intelligent Decision Technologies 2016* (Springer International Publishing, 2016), pp. 71–81

18. L. Breiman, J.H. Friedman, R.A. Olshen, C.J. Stone, *Classification and Regression Trees* (Chapman & Hall, New York, 1984)

19. M. Çelik, D. Karaboğa, F. Köylü, Artificial bee colony data miner (abc-miner), in *2011 International Symposium on Innovations in Intelligent Systems and Applications (INISTA)* (IEEE, 2011), pp. 96–100

20. B. Chai, X. Zhuang, Y. Zhao, J. Sklansky, Binary linear decision tree with genetic algorithm, in *13th International Conference on Pattern Rec., ICPR 1996, Vienna, Austria, 25–19 August, 1996* (1996), pp. 530–534

21. A. Chan, A.A. Freitas, A new ant colony algorithm for multi-label alssification with applications in bioinformatics, in *Proceedings of Genetic and Evolutionary Computation Conference (GECCO' 2006)* (San Francisco, 2006), pp. 27–34

22. C. Chen, Y. Chen, J. He, Neural network ensemble based ant colony classification rule mining, in *Proceedings of First International Conference Innovative Computing, Information and Control (ICICIC'06)* (2006), pp. 427–430

23. C. Chen, F. Ye, Particle swarm optimization algorithm and its application to clustering analysis, in *2012 Proceedings of 17th Conference on Electrical Power Distribution Networks (EPDC)* (IEEE, 2012), pp. 789–794

24. K. Chen, K. Wang, K.n Wang, M. Angelia, Applying particle swarm optimization-based decision tree classifier for cancer classification on gene expression data. Appl. Soft Comput. **24**, 773–780 (2014)

25. L. Chen, L. Tu, Parallel mining for classification rules with ant colony algorithm, in *Computational Intelligence and Security*, ed. by Y. Hao, J. Liu, Y.-P. Wang, Y.-M. Cheung, H. Yin, L. Jiao, J. Ma, Y.-C. Jiao. Lecture Notes in Computer Science, vol. 3801 (Springer, Berlin/Heidelberg, 2005), pp. 261–266

26. Y. Chiou, L.W. Lan, Genetic clustering algorithms. European J. Oper. Res. **135**(2), 413–427 (2001)

27. J. Chircop, C.D. Buckingham, The multiple pheromone ant clustering algorithm and its application to real world domains, in *2013 Federated Conference on Computer Science and Information Systems (FedCSIS)* (IEEE, 2013), pp. 27–34

28. P. Clark, R. Boswell. Rule induction with CN2: some recent improvements, in *Proceedings of European Working Session on Learning (EWSL-91)*. LNAI 482 (Springer, Berlin, 1991), pp. 151–163

29. M. Durgadevi, R. Kalpana, Medical distress prediction based on classification rule discovery using ant-miner algorithm, in *2017 11th International Conference on Intelligent Systems and Control (ISCO)* (IEEE, 2017), pp. 88–92

30. A. Esmin, R. Coelho, S. Matwin, A review on particle swarm optimization algorithm and its variants to clustering high-dimensional data. Artif. Intell. Rev. **44**(1), 23–45 (2015)

31. M.V. Fidelis, H.S. Lopes, A.A. Freitas, Discovering comprehensible classification rules with a genetic algorithm, in *Proceedings of the 2000 Congress on Evolutionary Computation, 2000*, vol. 1 (IEEE, 2000), pp. 805–810

32. G. Folino, C. Pizzuti, G. Spezzano, Genetic programming and simulated annealing: a hybrid method to evolve decision trees, in *Genetic Programming, European Conference, Edinburgh, Scotland, UK, April 15–16, 2000* (2000), pp. 294–303

33. A.A. Freitas, A genetic algorithm for generalized rule induction, in *Advances in Soft Computing* (Springer, 1999), pp. 340–353
34. A.A. Freitas, *Data Mining and Knowledge Discovery with Evolutionary Algorithms* (Media, Springer Science & Business, 2013), pp. 340–353
35. Z. Fu, B.L. Golden, S. Lele, S. Raghavan, E.A. Wasil, Diversification for better classification trees. Comput. OR **33**(11), 3185–3202 (2006)
36. M. Galea, Q. Shen, Simultaneous ant colony optimization algorithms for learning linguistic fuzzy rules, in *Swarm Intelligence in Data Mining*, ed. by A. Agraham, C. Grosan, V. Ramos (Springer, Berlin, 2006)
37. J. Handl, B. Meyer, Ant-based and swarm-based clustering. Swarm Intell. **1**(2), 95–113 (2007)
38. S. G. Heath, S. Kasif, S. Salzberg, Induction of oblique decision trees, in *Proceedings of the 13th International Joint Conference on Artificial Intelligence. Chambéry, France, August 28–September 3, 1993* (1993), pp. 1002–1007
39. N. Holden, A.A. Freitas, Web page classification with an ant colony algorithm, in *In Parallel Problem Solving from Nature—PPSN VIII*. LNCS 3242 (Springer, 2004), pp. 1092–1102
40. N. P. Holden, A.A. Freitas, A hybrid pso/aco algorithm for classification, in *GECCO '07: Proceedings of the 2007 GECCO Conference Companion on Genetic and Evolutionary Computation* (ACM, New York, NY, USA, 2007), pp. 2745–2750
41. N.P. Holden, A.A. Freitas, A hybrid pso/aco algorithm for discovering classification rules in data mining. J. Artif. Evol. Appl. **2008**, 2:1–2:11 (2008)
42. S. Hota, P. Satapathy, A.K. Jagadev, Modified ant colony optimization algorithm (mantminer) for classification rule mining, in *Intelligent Computing, Communication and Devices* (Springer, 2015), pp. 267–275
43. S. Izrailev, D. Agrafiotis, A novel method for building regression tree models for qsar based on artificial ant colony systems. J. Chem. Inform. Comput. Sci. **41**(1), 176–180 (2001)
44. D. Jankowski, K. Jackowski, Evolutionary algorithm for decision tree induction, in *IFIP International Conference on Computer Information Systems and Industrial Management* (Springer, 2014), pp. 23–32
45. W. Jiang, Y. Xu, Y. Xu, A novel data mining method based on ant colony algorithm, in *Advanced Data Mining and Applications*, ed. by X. Li, S. Wang, Z.Y. Dong. Lecture Notes in Computer Science, vol. 3584 (Springer, Berlin/Heidelberg, 2005), pp. 284–291
46. H. Kennedy, C. Chinniah, P.V.G. Bradbeer, L. Morss, The construction and evaluation of decision trees: a comparison of evolutionary and concept learning methods, in *Evolutionary Computing, AISB International Workshop, Manchester, UK, April 7–8, 1997* (1997), pp. 147–162
47. J. Kozak, U. Boryczka, Dynamic version of the acdt/acdf algorithm for h-bond data set analysis, in *ICCCI* (2013), pp. 701–710
48. J. Kozak, U. Boryczka, Goal-oriented requirements for acdt algorithms, in *International Conference on Computational Collective Intelligence* (Springer International Publishing, 2014), pp. 593–602
49. J. Kozak, U. Boryczka, Collective data mining in the ant colony decision tree approach. Inf. Sci. **372**, 126–147 (2016)
50. J. Kozak, P. Juszczuk, Association ACDT as a tool for discovering the financial data rules, in *IEEE International Conference on Innovations in Intelligent Systems and Applications, INISTA 2017, Gdynia, Poland, July 3–5, 2017* (2017), pp. 241–246
51. M. Kretowski, An evolutionary algorithm for oblique decision tree induction, in *International Conference on Artificial Intelligence and Soft Computing* (Springer, 2004), pp. 432–437
52. M. Kretowski, A memetic algorithm for global induction of decision trees, in *International Conference on Current Trends in Theory and Practice of Computer Science* (Springer, 2008), pp. 531–540
53. M. Kretowski, M. Grzes, Mixed decision trees: an evolutionary approach, in *International Conference on Data Warehousing and Knowledge Discovery* (Springer, 2006), pp. 260–269
54. M. Kretowski, M. Grzes, Evolutionary induction of mixed decision trees. Int. J. Data Warehous. Min. (IJDWM) **3**(4), 68–82 (2007)

55. S.S.S. Kumar, G. Divya, Data prediction and optimized clustering for mpso and cso based clustering. Pattern Int. J. Emerg. Technol. Eng. Res. (IJETER) **4**, 70–77 (2016)

56. R.J. Kuo, C.W. Shih, Association rule mining through the ant colony system for national health insurance research database in taiwan. Comput. Math. Appl. **54**, 1303–1318 (2007)

57. W. Kwedlo, M. Kretowski. An evolutionary algorithm using multivariate discretization for decision rule induction, in *European Conference on Principles of Data Mining and Knowledge Discovery* (Springer, 1999), pp. 392–397

58. B. Liu, H.A. Abbas, B. Mc Kay, Classification rule discovery with ant colony optimization. IEEE Comput. Intell. Bull. 31–35 (2004)

59. B. Liu, H.A. Abbass, B. Mckay, Density-based heuristic for rule discovery with ant-miner, in *6th Australasia-Japan Joint Workshop on Intelligent and Evolutionary Systems (AJWIS 2002)* (2002)

60. J.J. Liu, J.T. Kwok, An extended genetic rule induction algorithm, in *Proceedings of the 2000 Congress on Evolutionary Computation, 2000*, vol. 1 (IEEE, 2000), pp. 458–463

61. S. Madhusudhanan, M. Karnan, K. Rajivgandhi, Fuzzy based ant miner algorithm in datamining for hepatitis, in *International Conference on Signal Acquisition and Processing* (2010), pp. 229–232

62. D. Martens, M. De Backer, R. Haesen, B. Baesens, T. Holvoet, Ants constructing rule-based classifiers, in *Swarm Intelligence in Data Mining*, ed. by A. Agraham, C. Grosan, V. Ramos (Springer, Berlin, 2006)

63. D. Martens, M. De Backer, J. Vanthienen, M. Snoeck, B. Baesens, Classification with ant colony optimization. IEEE Trans. Evol. Comput. **11**, 651–665 (2007)

64. U. Maulik, S. Bandyopadhyay, Genetic algorithm-based clustering technique. Pattern Recogn. **33**(9), 1455–1465 (2000)

65. A. Niimi, E. Tazaki, Genetic programming combined with association rule algorithm for decision tree construction, in *Fourth International Conference on Knowledge-Based Intelligent Information Engineering Systems & Allied Technologies, KES 2000, Brighton, UK, 30 August–1 September 2000, 2 Volumes* (2000), pp. 746–749

66. T. Niknam, B. Amiri, An efficient hybrid approach based on pso, aco and k-means for cluster analysis. Appl. Soft Comput. **10**(1), 183–197 (2010)

67. M.P. Oakes, Ant colony optimization for stylometry: the federalist papers, in *Proceedings of Recent Advances in Soft Computing (RASC—2004)* (2004), pp. 86–91

68. F.E.B. Otero, A.A. Freitas, C. Johnson, cAnt-Miner: an ant colony classification algorithm to cope with continuous attributes, in *Ant Colony Optimization and Swarm Intelligence*, ed. by M. Dorigo, M. Birattari, C. Blum, C. Maurice, T. Stützle, A. Winfield. Lecture Notes in Computer Science, vol. 5217 (Springer, Berlin/Heidelberg, 2008), pp. 48–59

69. F.E.B. Otero, A.A. Freitas, C. Johnson, Handling continuous attributes in ant colony classification algorithms, in *CIDM* (2009), pp. 225–231

70. F.E.B. Otero, A.A. Freitas, C.G. Johnson, Inducing decision trees with an ant colony optimization algorithm. Appl. Soft Comput. **12**(11), 3615–3626 (2012)

71. R.S. Parpinelli, H.S. Lopes, A.A. Freitas, An ant colony algorithm for classification rule discovery, in *Data Mining: A Heuristic Approach*, ed. by H. Abbas, R. Sarker, C. Newton (Idea Group Publishing, London, 2002), pp. 191–208

72. R.S. Parpinelli, H.S. Lopes, A.A. Freitas, *Data Mining with an Ant Colony Optimization Algorithm* (IEEE Transactions on Evolutionary Computation, Special issue on Ant Colony Algorithms, 2004), pp. 321–332

73. J.R. Quinlan, *C4.5: Programs for Machine Learning* (Morgan Kaufmann, 1993)

74. K.R. Robbins, W. Zhang, J.K. Bertrand, R. Rekaya, The ant colony algorithm for feature selection in high-dimension gene expression data for disease classification. Math. Med. Biol. **24**, 413–426 (2007)

75. O. Roozmand, K. Zamanifar, Parallel ant miner 2, in *Artificial Intelligence and Soft Computing ICAISC 2008*, ed. by L. Rutkowski, R. Tadeusiewicz, L. Zadeh, J. Zurada. Lecture Notes in Computer Science, vol. 5097 (Springer, Berlin/Heidelberg, 2008), pp. 681–692

76. K. Salama, A. Abdelbar, Extensions to the ant-miner classification rule discovery algorithm, in *Swarm Intelligence*, ed. by M. Dorigo, M. Birattari, G. Di Caro, R. Doursat, A. Engelbrecht, D. Floreano, L.M. Gambardella, R. Gro, E. Sahin, H. Sayama, T. Stützle. Lecture Notes in Computer Science, vol. 6234 (Springer, Berlin/Heidelberg, 2010), pp. 167–178

77. K.M. Salama, A.M. Abdelbar, Using ant colony optimization to build cluster-based classification systems, in *International Conference on Swarm Intelligence* (Springer, 2016), pp. 210–222

78. K.M. Salama, A.M. Abdelbar, M. Ismail, Anwar, Data reduction for classification with ant colony algorithms. Intell. Data Anal. **20**(5), 1021–1059 (2016)

79. K.M. Salama, A.M. Abdelbar, A.A. Freitas, Multiple pheromone types and other extensions to the ant-miner classification rule discovery algorithm. Swarm Intell. **5**(3–4), 149–182 (2011)

80. K.M. Salama, A.M. Abdelbar, F.E.B. Otero, A.A. Freitas, Utilizing multiple pheromones in an ant-based algorithm for continuous-attribute classification rule discovery. Appl. Soft Comput. **13**(1), 667–675 (2013)

81. K.M. Salama, A.A. Freitas, Ant colony algorithms for constructing bayesian multi-net classifiers. Complexity **2**, 4 (2013)

82. K.M. Salama, A.A. Freitas, Clustering-based bayesian multi-net classifier construction with ant colony optimization, in *2013 IEEE Congress on Evolutionary Computation (CEC)* (IEEE, 2013), pp. 3079–3086

83. K.M. Salama, A.A. Freitas, Learning bayesian network classifiers using ant colony optimization. Swarm Intell. **7**(2–3), 229–254 (2013)

84. K.M. Salama, A.A. Freitas, Extending the abc-miner bayesian classification algorithm, in *Nature Inspired Cooperative Strategies for Optimization (NICSO 2013)* (Springer, 2014), pp. 1–12

85. K.M. Salama, F.E.B. Otero, Learning multi-tree classification models with ant colony optimization, in *IJCCI (ECTA)* (2014), pp. 38–48

86. G. Schaefer, Ant colony optimisation classification for gene expression data analysis, in *Rough Sets, Fuzzy Sets, Data Mining and Granular Computing*, ed. by H. Sakai, M. Chakraborty, A. Hassanien, D. Slezak, W. Zhu. Lecture Notes in Computer Science, vol. 5908 (Springer, Berlin/Heidelberg, 2009), pp. 463–469

87. G. Schaefer, Gene expression analysis based on ant colony optimisation classification. Int. J. Rough Sets Data Anal. (IJRSDA) **3**(3), 51–59 (2016)

88. P. Scheunders, A genetic c-means clustering algorithm applied to color image quantization. Pattern Recogn. **30**(6), 859–866 (1997)

89. P.S. Shelokar, V.K. Jayaraman, B.D. Kulkarni, An ant colony approach for clustering. Analytica Chimica Acta **509**(2), 187–195 (2004)

90. W. Siedlecki, J. Sklansky, A note on genetic algorithms for large-scale feature selection. Pattern Recogn. Lett. **10**(5), 335–347 (1989)

91. J. Smaldon, A.A. Freitas, A new version of the ant–miner algorithm discovering unordered rule sets, in *Proceedings of Genetic and Evolutionary Computation Conference (GECCO' 2006)* (San Francisco, 2006), pp. 43–50

92. G. Soliman, M. Khorshid, T. Abou-El-Enien, A hybrid ensemble classification algorithm using grey wolf optimizer for terrorism prediction. Int. J. Eng. Techn. Res. (IJETR) **5**, 183–190 (2016)

93. T. Stützle, H. Hoos, The MAX–MIN ant system and local search for the traveling salesman problem, in *Proceedings of IEEE–ICEC–EPS'97, IEEE International Conference on Evolutionary Computation and Evolutionary Programming Conference*, ed. by T. Baeck, Z. Michalewicz, X. Yao (IEEE Press, 1997), pp. 309–314

94. I. Surjandari, A. Dhini, A. Rachman, R. Novita, Estimation of dry docking duration using a numerical ant colony decision tree. Int. J. Appl. Manag. Sci. **7**(2), 164–175 (2015)

95. P. Tapkan, L. Özbakır, S. Kulluk, A. Baykasoğlu, A cost-sensitive classification algorithm: Bee-miner. Knowledge-Based Syst. **95**, 99–113 (2016)

96. K. Thangavel, P. Jaganathan, Rule mining algorithm with a new ant colony optimization algorithm. Int. Conf. Comput. Intell. Multimedia Appl. **2**, 135–140 (2007)

97. D.W. Van der Merwe, A.P. Engelbrecht, Data clustering using particle swarm optimization, in *CEC'03. The 2003 Congress on Evolutionary Computation, 2003*, vol. 1 (IEEE, 2003), pp. 215–220

98. P. Vergara, J.R. Villar, E. de la Cal, M. Menéndez, J. Sedano, Comparing aco approaches in epilepsy seizures, in *International Conference on Hybrid Artificial Intelligence Systems* (Springer, 2016), pp. 261–272

99. Z. Wang, B. Feng, Classification rule mining with an improved ant colony algorithm, in *Advances in Artificial Intelligence (Ai 2004)*. LNAI 3339 (Springer, Berlin, 2004), pp. 357–367

100. Y. Zhang, S. Wang, P. Phillips, G. Ji, Binary pso with mutation operator for feature selection using decision tree applied to spam detection. Knowledge-Based Syst. **64**, 22–31 (2014)

Chapter 3
Ant Colony Decision Tree Approach

3.1 Example

To start a detailed analysis of applying ant colony optimization to the problem of learning decision trees, we introduce a simple example. In this part of the chapter, we present an example of constructing a decision tree with the CART algorithm (all calculations are conducted based on the details given in Sect. 1.2). We also present the role of approximation algorithms (in this case, the ACO algorithm) in the process of learning decision trees.

We make use of the decision table presented in Chap. 1 (Table 1.1). On its basis, we introduce an example of constructing a decision tree, and the solutions that can be obtained using approximate algorithms. The solutions obtained when constructing decision trees with classical algorithms (in the form of decision trees seen as classifiers) are always the same. For example, the process of learning a decision tree based on the solutions known from the CART algorithm and described in Sect. 1.2 on the basis of Table 1.1 could eventually lead to the decision tree presented in Fig. 3.2—assuming some simplifications, like: treating the decision tree as knowledge representation (projection of a decision table) and constructing it without pruning.

The idea of constructing such a decision tree is relatively simple. First, we calculate (based on the decision table) the values of the splitting criteria for every node. Next, by selecting the highest value (the highest information gain), a data split is carried out based on the selected criterion. In this case, the splitting criterion selected was the Twoing criterion (1.13) with application of the membership test (1.8). After these calculations, the child nodes are generated. Data sets in those nodes differ, and the whole process is repeated. The first stage of decision tree construction is presented in Fig. 3.1. We should emphasize that as this is an example for the CART algorithm, a binary tree is eventually constructed.

For the given example, the criterion values are calculated for each possibility. Thus $(a_1, 0)$ (objects for which the value of attribute a_1 equals 0) are transferred to the left child node, and the remaining objects are transferred to the right child node.

© Springer International Publishing AG, part of Springer Nature 2019
J. Kozak, *Decision Tree and Ensemble Learning Based on Ant Colony Optimization*, Studies in Computational Intelligence 781,
https://doi.org/10.1007/978-3-319-93752-6_3

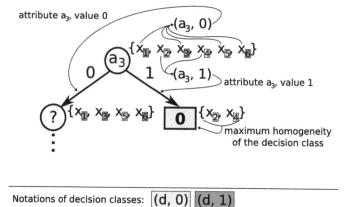

Notations of decision classes: (d, 0) (d, 1)

Fig. 3.1 Start of decision tree construction

It can be easily noticed that during the process of constructing the decision trees with the given applications in the root, attribute a_3 is selected for the value 0. This can be deduced from the calculations below (based on (1.14)):

$$(a_1, 0) = \frac{\frac{1}{3} \cdot \frac{1}{6}}{4} \cdot \left[\left| \frac{1}{3} - \frac{2}{3} \right| \cdot \left| \frac{2}{3} - \frac{1}{3} \right| \right]^2 = 0.0275$$

$$(a_2, 0) = \frac{\frac{2}{3} \cdot \frac{1}{3}}{4} \cdot \left[\left| \frac{1}{2} - \frac{1}{2} \right| \cdot \left| \frac{1}{2} - \frac{1}{2} \right| \right]^2 = 0.0556$$

$$(a_3, 0) = \frac{\frac{2}{3} \cdot \frac{1}{3}}{4} \cdot \left[\left| \frac{1}{4} - \frac{3}{4} \right| \cdot \left| \frac{0}{2} - \frac{2}{2} \right| \right]^2 = 0.1251$$

$$(a_4, 0) = \frac{\frac{1}{3} \cdot \frac{1}{3}}{4} \cdot \left[\left| \frac{2}{3} - \frac{1}{3} \right| \cdot \left| \frac{1}{3} - \frac{2}{3} \right| \right]^2 = 0.0275$$

After the division, two child nodes are generated, according to the values of attribute a_3. To the left node, objects x_1, x_3, x_5 and x_6 are assigned, and to the right node—objects x_2 and x_4. Now, the same process is repeated in the left node, so further divisions can be made. As in the right node the maximal homogeneity of the decision class is acquired (all objects are in class 0), the node is treated as a leaf with label 0. This stage of the algorithm is presented in Fig. 3.1.

The complete decision tree is presented in Fig. 3.2. In case of classical algorithms, the final decision tree will always be exactly the same for a given algorithm—but it can be different, say, when using C4.5 [10] from the tree obtained with CART [7]).

Use of approximate algorithms allows for searching a higher number of solutions, and reinforcement learning allows for additional improvement of those solutions.

Fig. 3.2 Example decision
tree for decision Table 1.1

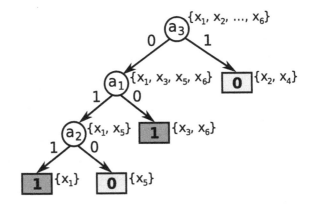

Bearing in mind the modus operandi of the ant colony optimization algorithm introduced in the example described in Sect. 1.3.2 and the example given above, we may assume that in case of using a split criterion (similarly as above) as the heuristic function in the ant algorithms, the obtained decision trees can be improved. In the first phases of the algorithm operation, the generated decision trees are similar to those that would be learned on the basis of a specific split criterion—for random changes are introduced (according to the pseudo random proportionality rule (1.19)). Next, if the introduced changes lead to obtaining better results, they are remembered as the pheromone trail, and eventually recreated and improved again.

An example of such a solution for the analyzed decision table (Table 1.1) is introduced in Fig. 3.3. It shows in part (a) the decision tree generated as a result of the first iteration of the ant algorithm (pheromone trail values are given in parentheses). In the subsequent iterations, the decision tree is modified further. This is why an algorithm run results in obtaining different decision trees—for example, the final tree is shown in part (b) of Fig. 3.3.

As we may see, in this particular case a stochastic algorithm accompanied by a heuristic based on the Twoing criterion has allowed for creating a decision tree with a better structure than that yielded by the classical algorithm. Of course, this is only an example, and the constructed decision tree has not been tested in classification. However, this example allows us to show the idea of the ACDT algorithm. The exact solutions and their justifications used in ACDT can be found further on in this chapter.

3.2 Definition of the Ant Colony Decision Tree Approach

The ACDT algorithm is an ACO algorithm used to construct decision trees. As a result of running the algorithm, a classifier, i.e., a decision tree, is created. This algorithm is nondeterministic; therefore, each execution of the algorithm typically leads to the construction of a different decision tree. The principle of ACDT is based

Table 3.1 Original parameters in data sets

Data set	No. of instances	No. of att.	Dec. class	Data set	No. of instances	No. of att.	Dec. class
australian	690	14	2	imports-85	205	25	6
balance-scale	625	4	3	ionosphere	351	34	2
breast-cancer	280	9	2	iris	150	4	3
breast-tissue	106	9	6	jsbach	5665	16	102
car	1728	6	4	kr-vs-kp	3196	36	3
cleveland	303	13	5	mushroom	8124	22	7
CTG	2126	35	10	nursery	12960	8	5
dermatology	366	34	6	optdigits	5620	64	10
ecoli	336	7	8	parkinsons	195	22	2
glass	214	9	6	shuttle	72500	9	7
heart	270	13	2	soybean	307	35	19
hepatitis	155	19	2	transfusion	722	4	2
horse-colic	366	22	2	wine	178	13	3
house-votes-84	425	16	2	yeast	1484	9	10
hungarian	293	13	5	zoo	92	16	7

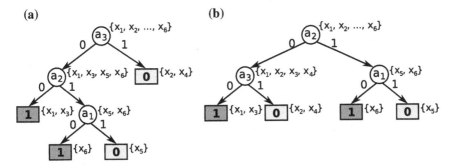

Fig. 3.3 Example decision trees for decision Table 1.1 obtained using the ACO approach

on using a pheromone trail that is laid on the edges, and on heuristics that are used in classical algorithms.

The ACDT approach can be represented as a five-tuple:

$$ACDT = \left\langle (X, A \cup \{c\}), T(S), ants, p_{m, m_{L(i,j)}}(t), S \right\rangle,$$ (3.1)

which contains:

$(X, A \cup \{c\})$—decision table which represents the problem and is expressed as an ordered pair, where X is a set of objects and A a set of attributes, including a decision attribute c);

$T(S)$—decision tree, which is a solution to the problem;

As well as the following elements of the ant colony system:

$ants$—number of ants in an iteration,

$p_{m, m_{L(i,j)}}(t)$—solution selection rule for each node of the decision tree at time t,

S—set of acceptable objects in a node that indirectly performs the function of the taboo list.

The decision tree can represent any hypothesis that is acceptable for a given set of attributes. This means that for a decision tree representing the hypothesis that the data set X consists of n data items (as in formula (1.1))

$$X = \{x_1, x_2, \ldots, x_n\},$$ (3.2)

described by m attributes

$$A = \{a_1, a_2, \ldots, a_m\},$$ (3.3)

and belonging to one of C decision classes, each object in the set X can be described (based on (1.3)) as:

$$x_i = ([v_i^1, \ldots, v_i^m], c_i), v_i^j \in A_j, c_i \in \{1, \ldots, C\},$$ (3.4)

where v_i^j is the value of attribute a_j for object x_i, and c_i is the decision class for object x_i.

Each node of the decision tree contains a test on attributes

$$te: X \rightarrow R_{te}, \qquad (3.5)$$

where $R_{te} = \{r_1, r_2, \ldots, r_z\}$ is the set of possible tests for a node.

If we know that the test is applied to values of attribute $a: X \rightarrow A$, then we write

$$te: A \rightarrow R_{te}. \qquad (3.6)$$

Assume test r_i creates T_i subtrees. Then for each node tests $r_1, r_2, \ldots, r_z$ build $T_1, T_2, \ldots, T_z$ possible subtrees. Therefore, the hypothesis represented by the node can be written as $(\forall x \in X)h(x)$, where:

$$h(x) = \begin{cases} h_1(x), \ te(x) = r_1 \\ h_2(x), \ te(x) = r_2 \\ \vdots \qquad \vdots \\ h_z(x), \ te(x) = r_z. \end{cases} \qquad (3.7)$$

In contrast, the decision tree $T(S)$ can be used for classification based on the following formula (where D denotes distribution):

$$\epsilon(T(S), D) = \sum_{(x,y) \in U} D(x, y) \cdot L(y, T(S)(x)), \qquad (3.8)$$

with $L(y, T(S)(x))$ being the function:

$$L(y, T(S)(x)) = \begin{cases} 0, \ if \ y = T(S)(x) \\ 1, \ if \ y \neq T(S)(x), \end{cases} \qquad (3.9)$$

where:

$T(S)$—decision tree T constructed based on training set S,

$T(S)(x)$—decision for object x determined by its conditional attributes,

U—set of possible values for each attribute.

The evaluation function for decision trees is calculated based on the following formula:

$$Q(T) = \phi \cdot s(T) + \psi \cdot a(T, S), \qquad (3.10)$$

where:

$s(T)$—size of decision tree T,

$a(T, S)$—accuracy of classification of an object from a clean set (also called the validation set) S by tree T,

ϕ and ψ—constants that define relative importance of $s(T)$ and $a(T, S)$.

The size of decision tree T is calculated as follows:

$$s(T) = \frac{1}{n},$$

(3.11)

where n is the number of nodes.

The accuracy of classification is calculated as follows:

$$a(T, S) = \frac{\sum_{c=1}^{C}(TP_c)}{|S|},$$

(3.12)

where:

TP_c—number of correctly classified objects in class c,

$|S|$—number of all objects in the clean set,

C—number of decision classes.

The probability of choosing the appropriate test in the node is calculated using the formula

$$p_{m,m_{L(i,j)}}(t) = \frac{\tau_{m,m_{L(i,j)}}(t) \cdot \eta_{i,j}^{\beta}}{\sum_{i}^{a} \sum_{j}^{b_i} \tau_{m,m_{L(i,j)}}(t) \cdot \eta_{i,j}^{\beta}},$$

(3.13)

where:

m can be interpreted as $m_{p,o}$—that is, the superior node at which the test for attribute p and value o is performed.

$\eta_{i,j}$—heuristic value for the test concerning attribute i and value j,

$\tau_{m,m_{L(i,j)}}$—amount of pheromone available at time t on the connection between nodes m and m_L (concerning attribute i and value j),

β—relative importance of the heuristic, experimentally determined as equal 3.

In the case of ACDT, the value of the heuristic function for each attribute-value pair at node m is as follows:

$$\forall_{i \in A, j \in V_i} \quad \eta_{i,j} = \frac{P_l P_r}{4} \left[\sum_{c=1}^{C} |p(c|m_l) - p(c|m_r)| \right]^2,$$

(3.14)

where:

V_i—set of values of attribute i,

$p(c|m_l)$—conditional probability of class c provided at node m_l,

P_l / P_r—probability of object transition into the left / right node: m_l or m_r, respectively

a_j—j-th variable,

a_j^R—the best splitting value of variable a_j.

ACDT, which handles continuous attributes, incorporates an inequality test based on (1.7):

$$T(x) = \begin{cases} 1, & \text{if } a_j(x) \le cut \\ 0, & \text{otherwise } (a_j(x) > cut) \end{cases},$$ (3.15)

where cut is a threshold value (cut-off point).

The effects of these changes influence the precision and prediction of splitting continuous values. The observations are as follows: cut-off points are selected based on the values occurring for the analyzed attributes in the testing set. Those points can be assessed empirically in a series of experiments with a substantial number of learning tasks. For each attribute, the cut-off point is obtained for specific values from the testing set. Then, it is evaluated.

Please note that in ACDT the choice of the splitting rule at an appropriate node is performed based on the random-proportional rule (3.13) only when $q > q_0$. We create the formula

$$r = \begin{cases} \arg \max_{m_L \in R_m^c} \{[\tau_{m,m_L}(t)] \cdot [\eta_{i,j}]^\beta\} & \text{if } q \le q_0 \\ J & \text{if } q > q_0 \end{cases},$$ (3.16)

where J is randomly chosen based on the probability $p_{m,m_{L(i,j)}}(t)$ (Eq. (3.13)).

This random choice means that when $q \le q_0$, greedy approach is used as a consequence of the maximum value of the heuristic-pheromone product (see (3.14)).

The next problem worth further study is pheromone increase. Edges of the decision tree have pheromone values associated to them. The initial value of the pheromone trail is established based on the number of attribute values (see (3.17)).

$$\tau_{m,m_L}(t = 0) = \frac{\log_2(C)}{\sum_{att=1}^{|A|} |a_{att}|},$$ (3.17)

where:

$|A|$—number of attributes,

$|a_{att}|$—number of possible values of attribute a_{att}.

All possible combinations of the nodes (edges) of the decision tree are stored in a pheromone matrix. Pheromone updates are performed (3.18) by increasing the previous value for each parent–child pair of nodes as follows:

$$\Delta\tau_{m,m_L}(t + 1) = (1 - \gamma) \cdot \tau_{m,m_L}(t) + Q(T),$$ (3.18)

where $Q(T)$ denotes the evaluation function of the decision tree (see (3.10)), and γ is a parameter representing the evaporation rate, which usually equals 0.1.

Decision tree construction is an iterative decision-making process that involves selecting splits for each node. In ACDT, the selection of splits depends on the existing decision tree structure (an attribute and its acceptable values in the superordinate, parent node). As an example, the core principle of the algorithm is that a split in

(a)

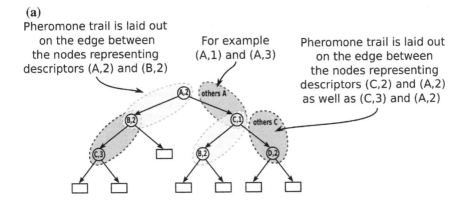

(b)

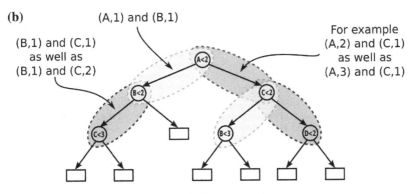

Fig. 3.4 Pheromone trail deposition in ACDT

a node is selected, and therefore, an edge is created. Then, a pheromone trail can be laid out on this edge. In consequence, the updated pheromone trail for an edge connecting attributes $(Att_A, Val_1, Att_B, Val_0)$ does not have the same value as for the edge with the opposite order of the attribute sequence $(Att_B, Val_0, Att_A, Val_1)$. An example of how the pheromone trail is laid in ACDT is shown in Figs. 3.4 and 3.5.

The process of constructing a decision tree using ACDT is presented in Algorithm 2. After the initial value of the pheromone trail is established, each iteration of ant agents constructs a decision tree, as shown in lines 2–23. Each ant agent in an iteration constructs a decision tree (lines 4–18) by taking into account the values of the heuristic function and of the pheromone trail determined when each of the nodes is analyzed (line 11). The ant agent that has constructed the best decision tree in the iteration (in terms of quality evaluation (3.10)) lays a pheromone trail. The value of the pheromone laid depends on the value of $Q(T)$ (lines 15–17 and 19). The best

Fig. 3.5 Example of ACDT
operation on the pheromone
trail

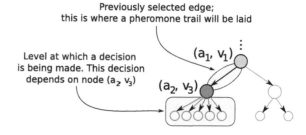

decision tree, chosen by means of evaluation carried out throughout the algorithm
operation, is returned by the algorithm as its result.

Algorithm 2: Pseudo-code of the ACDT algorithm

```
1   pheromone = initialization_pheromone_trail(); // by Eq. (3.17)
2   for number_of_iterations do
3     best_tree = NULL;
4     for number_of_ants do
5       //build the decision tree
6       new_tree = null;
7       while (stop_condition_is_not_fulfilled)
8         heuristic = calculate_the_heuristic_function(); // by Eq. (3.14)
9         p = calc_the_choosing_probability(pheromone, heuristic); // by Eq. (3.13)
10        //choose the test in the node (roulette wheel)
11        new_tree→test = roulette_wheel(p_{m,m_{L(i,j)}}(t));
12      endWhile
13      pruning(new_tree);
14      assessment_of_the_tree_quality(new_tree); // by Eq. (3.10)
15      if new_tree is_higher_quality_than best_tree then
16        best_tree = new_tree;
17      endIf
18    endFor
19    update_pheromone_trail(best_tree, pheromone); // by Eq. (3.18)
20    if best_tree is_higher_quality_than best_constructed_tree then
21      best_constructed_tree = best_tree;
22    endIf
23  endFor
24  result = best_constructed_tree;
```

In the case of ACDT, top-down induction of the decision trees is analyzed. In
each node of the constructed decision tree, a specific test is stored for the analyzed
attribute. During the classification of objects, an object that has passed this test moves
to the left descendent node; in the opposite case, it moves to the right descendent
node. Decision trees constructed by ACDT provide 100% coverage.

This approach, called error-based pruning (which is implemented in the C4.5
algorithm), considers each of the decision nodes in the tree as a candidate for pruning.

Pruning of a decision node consists in removing the subtree rooted at the node, thus making it into a leaf node, and assigning it to the most common classification of the training examples associated with that node [10, 11]. Nodes are removed only if they meet the following condition:

$$\epsilon'(prun(T, m), S) \leq \epsilon'(T, S) + \sqrt{\frac{\epsilon'(T, S) \cdot (1 - \epsilon'(T, S))}{|S|}}, \qquad (3.19)$$

where:

T—decision subtree whose root is a candidate node,
S—training set,
m—candidate node in the decision tree,
$prun(DT, m)$—tree with node m that is replaced with a leaf.

Construction of the decision tree may be stopped in the following cases:

- the data set is empty,
- all objects belong to the same decision class,
- there is no test that could result in further divisions.

Thus, the following stopping criterion can be applied:

$$S = \emptyset \vee |\{c' \in C | (\exists x \in S)dec(x) = c'\}| = 1 \vee D = \emptyset, \qquad (3.20)$$

where c is a decision class value.

In both of the cases described above, the decision for the leaf is determined in accordance with the majority rule, i.e., a correct selection is made for most objects. When the data set is empty, the decision is chosen for the majority of objects from the learning set:

$$label(S, c) = \begin{cases} c_{max}, & \text{if } S = \emptyset \\ \arg\max_{c \in C} |S_c|, & \text{otherwise,} \end{cases} \qquad (3.21)$$

where C is the set of decision class values that determines the value of the decision class which occurs most often in the original learning set.

3.3 Machine Learning by Pheromone

Reinforcement learning involves problem solving using a reward that grows for good solutions during the algorithm run. The idea of reinforcement learning is based on behaviorism, which is a psychological approach assuming that behavior of an individual depends both on the individual's environment and on the system of rewards and punishments that are used to control the behavior of that individual. The main difference between classical techniques and reinforcement learning algorithms is that in the latter case there is no need for information about the Markov decision process [1].

In the ACDT algorithm, pheromone trail update acts as a system of rewards and punishments. Pheromone trail evaporation means a punishment for those edges that have not been selected, and pheromone trail laying means a reward for those edges that have been selected. This allows the ant agents to adapt to the environment, i.e., the solution space that depends on the analyzed data set.

We may say that in this particular case the punishment and reward system is defined with the formula (3.18), where reinforcement of the pheromone trail is denoted as a reward. The reward is implemented by adding the value $Q(T)$ corresponding to the quality of the constructed decision tree (3.10). In turn, evaporation of the pheromone trail corresponds to punishment—it takes place for every edge (both those that occur in the current decision tree and those that do not). The value of the punishment depends directly on the parameter γ—a higher value of this parameter leads to faster pheromone evaporation.

It should be noted that in the analyzed case the value of the pheromone trail is mapped in time (which is not always the case—for example, in case of the Ant Miner algorithm there is no time-based distinction). More precisely, occurrence of a given edge in the initial phase of the algorithm is less important than its occurrence in the final iterations. This is connected with the increasing strength of algorithm adaptation.

For example, if the initial value of the pheromone trail is 0.5 for all edges, and the parameter γ equals 0.1, then at algorithm start the example edges will have the following values: $((a_1, 0), (a_3, 0), 0.5)$, $((a_1, 0), (a_2, 0), 0.5)$, $((a_1, 0), (a_3, 1), 0.5)$ and $((a_1, 0), (a_4, 0), 0.5)$—which is shown in Fig. 3.6. It should be noted that the values have been artificially prepared for the purpose of explaining the proposed solution. Obviously, in the proposed system the number of possible edges is much greater (for example, for 4 attributes with binary values there would be initially 12 possible combinations—assuming that a binary decision tree for two attribute values can be treated as symmetrical).

If at the next step the constructed decision tree contains the edges $((a_1, 0), (a_3, 0))$ and $((a_1, 0), (a_3, 1))$, and the quality of that tree is, say, $Q(T) = 0.8$, the edges in the pheromone map will be as follows: $((a_1, 0), (a_3, 0), 1.17)$, $((a_1, 0), (a_2, 0), 0.45)$, $((a_1, 0), (a_3, 1), 1.17)$ and $((a_1, 0), (a_4, 0), 0.45)$. If the decision tree in the next iteration contains the edges $((a_1, 0), (a_3, 0))$ and $((a_1, 0), (a_4, 0))$ and its quality is once more equal $Q(T) = 0.8$, the edges will look as follows: $((a_1, 0), (a_3, 0), 1.773)$, $((a_1, 0), (a_2, 0), 0.405)$, $((a_1, 0), (a_3, 1), 1.053)$ and $((a_1, 0), (a_4, 0), 1.125)$. We can see that in such a trivial case after two algorithm iterations two edges which were once included in the previously constructed best decision trees (with the same quality) have different values of the pheromone trail. Edge $((a_1, 0), (a_3, 1), 1.053)$, which occurred earlier, has a smaller value of the pheromone trail (thus a lower support value) than edge $((a_1, 0), (a_4, 0), 1.125)$, which occurred later.

Pheromone maps represent the distribution of pheromone trails that are laid by ant agents in the solution space. These maps illustrate graphically the pheromone

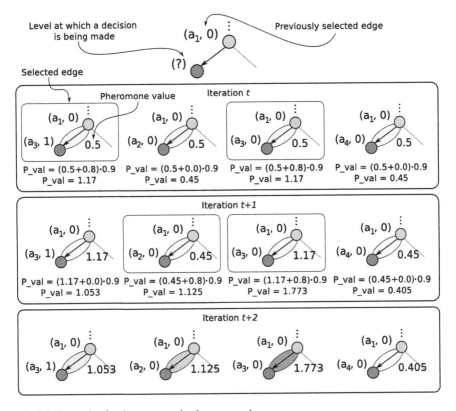

Fig. 3.6 Example of a pheromone value between nodes

trail values that are assigned to a particular solution—the greater the pheromone trail value, the more ant agents have chosen a given solution. For the ACDT algorithm, in which the pheromone trail is laid on the edges, pheromone trail values are assigned to edges between parent-child pairs of nodes.

Pheromone maps determine the values for all possible edges in the decision tree. The parent axis is responsible for the parent node, i.e., the node that already exists in the tree under construction. For this node, a test on an attribute is defined (a split in the tree). In the analyzed case, the child axis represents a child node (selected by an ant agent) which is dependent on the parent node. The value axis presents the pheromone trail value for the parent-child edge. Nodes are represented by descriptors (attribute-value pairs), whereas the pheromone trail matrix records all possible descriptors, i.e., all attributes and their values, both for the parent and the child nodes.

Figure 3.7 presents an example of a pheromone map with the relevant description. A higher pheromone value for a pair of descriptors means stronger reinforcement for selecting such an edge in the constructed decision tree. In this solution, only edges are recorded (thus the tree level is not recorded). This ensures greater possibility of

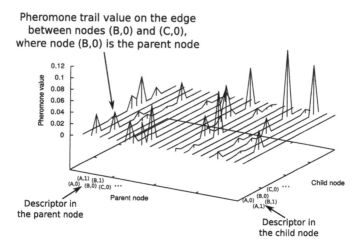

Fig. 3.7 Example of a pheromone map created when ACDT finished running

modifying the initial tree structure, because for a given descriptor selection of the child node is not limited by the current level of the decision tree.

For simplicity, we present the pheromone map in case of using the equality test (1.6). If the membership test is used (1.8), all possible combinations of descriptors for the given case should be considered. In case of using the inequality test (1.7), one should proceed in a similar way.

Experimental analysis of pheromone maps during the algorithm run is described in detail in Sect. 3.4.1. However, we should mention here that the performed analysis allows for observing certain characteristics. The dynamically (but gradually, not suddenly) changing pheromone maps imply that the process of learning decision trees with the use of ant colony optimization algorithms is not—despite the stochastic applications—a random walk, but a gradual adaptation. This is true even in the case when initially bad solutions are transformed into decision trees of a visibly better quality. Moreover, regardless of the initial values of pheromone maps, in the final phases of the algorithm operation, certain characteristic solution landscapes can be noticed (the same for a given problem).

3.4 Computational Experiments

A range of experiments have been conducted to test cooperation between ant agents in ACDT. First, we describe our experimental methodology and explain the motivation behind it. Then, we present and discuss our results. In this chapter we consider an experimental study carried out with the following parameters. We conducted 30 experiments for each data set, whereby each experiment included 900 executions with the ant colony size of 15. A comparative study of the ACDT algorithm with four different versions of quality assessment was also carried out.

Evaluation of the ACDT approach was conducted using 30 public-domain data sets from the UCI (University of California at Irvine) data set repository. Table 3.1 shows the main characteristics of the data sets. All experiments were conducted with (3.10) parameters of $\psi = 1.0$ and $\phi = 1.0$. The parameter values employed in ACO were established using the method first presented in [2], i.e., $q_0 = 0.2$, $\beta = 3.0$, $\gamma = 0.1$.

Data sets larger than 1000 samples were randomly divided into three groups, namely the training, testing and cleaning sets, respectively. Data sets with fewer than 1000 samples were estimated using 10-fold cross-validation. We used an additional data set, called a clean set (or a validation set). The results were examined on a clean set that had not been used to build a new classifier.

We assumed that all experiments should be conducted in the same conditions. The main aspect of this chapter was correlated with analysis of ACDT, which determined such a solution. However, during the comparison with other algorithms, the standard rules were preserved.

With such assumptions, every data set described in the Table 3.1 was prepared according to the rules given below. All attributes were unified to a numeric range in such a way that all data items were sorted (in an increasing order, and in lexicographic order in case of nominal data—though such data was analyzed classically). In the next step, all sorted values were assigned consecutive numbers. Such a dictionary (for sorted data) was used to pre-process the real world data—now without sorting. This enabled maintaining the same format in all files. Moreover, this also allowed for dividing the data in the same manner: for learning and for testing (for every experiment). Most of the results presented in this chapter were obtained based on p data sets reprocessed in this way. This allowed us to analyze observations of ant algorithms' operation in the process of constructing decision trees. Only the comparison with other algorithms (Sect. 3.4.3) was made without any preprocessing, which could be considered as an additional handicap for the algorithm. However, it should be pointed out that preprocessed data sets are more difficult to use for building classifiers in case of attributes being floating point real numbers. This may affect the quality of classification compared to that based on unprocessed data.

Processing of the data did not affect the movement of objects between the decision classes, the number of classes, etc.,—everything was preserved. Processing, however, had influence on data sets with attributes being real numbers. In such a situation, we can divide the analyzed data sets into three groups of similar size (Table 3.2):

1. Data sets where processing had no impact—"attributes unchanged".
2. Data sets where processing had impact on some attributes (usually less than 50%)—"attributes partly changed".
3. Data sets where processing had impact on most of the attributes)—"attributes changed".

The motivation behind the experiments discussed in this chapter was our intent to explain the universality of ACDT (which can yield good results even without data

Table 3.2 Three groups of data sets

Group 1 contains 11 data sets ("attributes unchanged")	Group 2 contains 9 data sets ("attributes partly changed")	Group 3 contains 10 data sets ("attributes changed")
balance-scale	australian	breast-tissue
breast-cancer	cleveland	ecoli
ear	CTG	glass
dermatology	heart	ionosphere
house-votes-84	hepatitis	iris
jsbach	horse-colic	parkinsons
kr-vs-kp	hungarian	shuttle
mushroom	imports-85	transfusion
nursery	optdigits	wine
soybean		yeast
zoo		

cleaning [5] or tuning the parameters), as well as the preliminary tests on cooperation described in [3]. The aim of these experiments was to test the algorithm by applying the pheromone trail and the heuristic, and to examine their importance for the algorithm. The results of experiments were to explain what is really behind the good results produced by ACDT. These experiments will also allow us to determine the role of ant colony optimization algorithms in the process of constructing a decision tree. This can be achieved by analyzing how the algorithm performs without additional heuristic knowledge, i.e. based merely on the ant agents' capability of learning through the use of pheromone trail.

Comparison of ACDT, the heuristics used and the pheromone trail, as well as analysis of the pheromone maps determined in the subsequent stages of the algorithm operation, will allow us to observe important relationships. The experiments will allow for evaluating the effectiveness of ACDT. This is an algorithm that enables more precise analysis of the solution space during decision tree construction. In consequence, many local optima can be tested, and global optima can possibly also be found.

3.4.1   Reinforcement Learning via Pheromone Maps

The results of experiments aimed at confirming that the considered algorithm learns based on the pheromone trail are shown in Table 3.3. The table presents the classification accuracy and the number of decision tree nodes for the algorithm using the pheromone trail. More precisely, we present the result obtained after the algorithm has finished running, as well as for the decision tree created after the first iteration of the algorithm. This methodology has been suggested for comparing the

Table 3.3 Comparison of the decision tree constructed by the first iteration with the best decision tree obtained after 60 iterations

Data set	Pheromone only		First iteration		Without heuristic f. and pheromone	
	Accuracy	Num. of nodes	Accuracy	Num. of nodes	Accuracy	Num. of nodes
australian	0.7913	37.5	0.5872	8.0	0.6149	13.8
balance-scale	0.8102	58.8	0.7546	57.1	0.7654	57.2
breast-cancer	0.6996	14.8	0.6723	3.6	0.6944	4.8
breast-tissue	0.4131	11.0	0.2806	3.5	0.2980	4.7
car	0.9046	65.2	0.7325	45.5	0.7444	47.9
cleveland	0.5416	14.2	0.5278	4.4	0.5343	6.8
CTG	0.6771	95.6	0.2831	10.3	0.3026	19.5
dermatology	0.8705	24.8	0.4362	7.6	0.4992	10.4
ecoli	0.7781	20.2	0.5525	7.3	0.6258	13.4
glass	0.5996	19.4	0.4144	4.7	0.4594	7.8
heart	0.7564	20.5	0.6049	4.6	0.6269	7.9
hepatitis	0.7982	9.9	0.7867	3.9	0.7917	4.5
horse-colic	0.8200	15.0	0.7411	8.2	0.7504	9.0
house-votes-84	0.9281	7.5	0.7271	8.8	0.7488	8.8
hungarian	0.6301	7.0	0.6243	3.8	0.6299	5.4
imports-85	0.5602	23.6	0.4108	6.8	0.4136	7.9
ionosphere	0.8556	15.2	0.7299	7.6	0.7410	11.8
iris	0.9252	6.8	0.7107	5.8	0.8103	7.6
jsbach	0.6375	260.0	0.1179	22.3	0.1238	27.4
kr-vs-kp	0.9105	98.9	0.5926	25.6	0.6289	39.0
mushroom	0.6295	153.7	0.5456	98.8	0.5623	117.8
nursery	0.9314	347.5	0.5979	287.4	0.6168	265.9
optdigits	0.6609	307.0	0.4505	162.8	0.4972	203.3
parkinsons	0.8358	11.9	0.7485	4.6	0.7709	6.9
shuttle	0.9826	412.2	0.8041	118.9	0.8502	133.3
soybean	0.7252	38.5	0.3783	16.0	0.4271	17.2
transfusion	0.7134	2.5	0.7039	6.0	0.7114	9.9
wine	0.8419	12.3	0.6059	7.0	0.6613	8.9
yeast	0.4665	52.3	0.3451	5.3	0.3649	15.5
zoo	0.8081	8.0	0.6220	6.2	0.6323	6.3

results obtained by the algorithm with and without the pheromone trail. The above information is crucial to understanding the way in which learning takes place via collaboration between ants; it allows us to demonstrate the quality improvement due to ants that have used the pheromone trail and contributed to swarm intelligence.

Fig. 3.8 Convergence chart of ACDT for the *nursery* base example

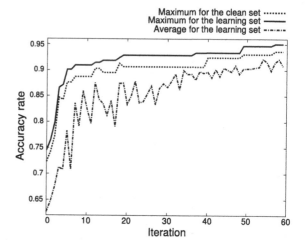

The results obtained confirm that the algorithm can learn through communication between ant agents via the pheromone trail. In each case, the classification accuracy is improved compared to the tree constructed by the first iteration of ant agents. A significant improvement in the classification accuracy (30–55%) can be observed for the dermatology, soybean and five other data sets, i.e. for large data sets that most data classification algorithms find difficult to analyze. A large difference in the classification accuracy (15–25%) can also be observed for eleven subsequent data sets. For seven data sets, the results are improved by 5–15%, whereas for five other data sets (specifically, small data sets, which are easy to classify in terms of structure and complexity) the results are improved by less than 5%. This phenomenon is probably caused by the characteristics of the solution space, where the optimal solutions are either very easy or very difficult to find.

We have also compared our results to random search, where the number of iterations is equal to (*number of iterations · number of ants*) in the ACDT approach. The random search approach performs worse than the ACDT approach with pheromone updating only. The pheromone values are changed in the course of applying our approach, which constructs a representation of the "colony knowledge".

To examine the performance of ACDT, convergence of that algorithm in case of considering the pheromone trail has been analyzed. Figure 3.8 shows convergence charts for a selected data set (for better visualization, the most interesting case has been chosen). The algorithm operates similarly to other typical ACO implementations (in various applications). After a rapid improvement at the beginning, a stagnation phase follows. Finally, a slight upturn of the results is observed. The stagnation phase occurs between 30–45 iterations, which is 60% of the running time (with predefined values of ACO parameters).

Figure 3.9 presents pheromone trail values after completion of the algorithm run—namely, six different examples for the *balance − scale*. In turn, Figs. 3.10 and 3.11 show pheromone maps at different stages of the algorithm operation (i.e., after 10,

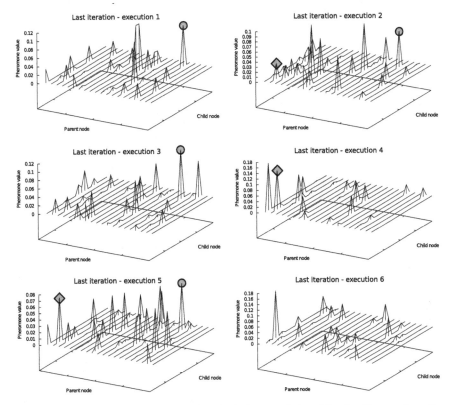

Fig. 3.9 Pheromone maps (pheromone trail values) after completion of the algorithm run (i.e., after 60 iterations) for the *balance − scale* base example

20, 30, 40, 50, and 60 iterations). For all of the analyzed data sets, the charts for pheromone maps show similar relationships. However, the *balance − scale* base was chosen for presenting as an example because the small numbers of attributes and their values in that base make the charts more readable.

As shown in Fig. 3.9, independent runs of the algorithm for the same data sets (and for the same parameter values and experimental conditions) may each time produce different pheromone maps, and therefore construct decision trees with different edges. However, as the differences in classification quality and decision tree size are small, the generated trees are different, but of the same quality. Certain edges (parent-child attribute-value pairs) can, however, be observed in the majority of pheromone maps (shown in Fig. 3.9). This proves that some edges are repeatedly selected by ant agents during independent runs of the algorithm. The above observation can be used to analyze the relationships between the attributes and their values during attribute selection.

The rhombuses (diamonds) and circles (bubbles) shown in Fig. 3.9 can represent memes that are producing the relation between the attributes investigated in a given

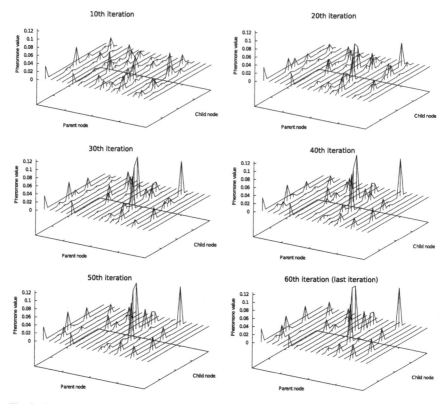

Fig. 3.10 Pheromone maps (1) (pheromone trail values) after 10, 20, 30, 40, and 60 iterations for the *balance − scale* base example

data set. In the further development of this algorithm, this property allows for making use of the knowledge represented by an artificial ant colony.

Figures 3.10 and 3.11 show pheromone maps that are visualized at different stages of the algorithm operation. The images present two different runs of the algorithm. In both cases, the connections between the nodes (edges) that were assigned large pheromone trail values at the initial stages of the algorithm operation often have minimum values at the end (as a result of pheromone trail evaporation). The edges that are selected repeatedly are established only after approximately 20–30 iterations; however, the pheromone map that is determined when the algorithm has finished running is nonetheless different from the maps that were determined after 20 or 30 iterations.

In fact, ACDT involves search of a space that initially resembles exploration of the solution space (with many pairs of descriptors and a pheromone trail that has been laid on them). Later, this space is gradually exploited. The moment of entering the exploitation phase can be seen in the chart showing the pheromone map after 30 iterations of the ant agents. At that point, the pheromone trail values usually start

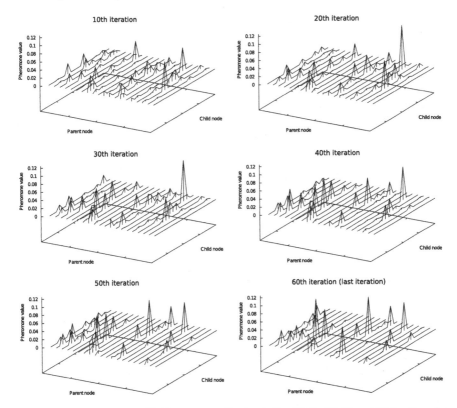

Fig. 3.11 Pheromone maps (2) (pheromone trail values) after 10, 20, 30, 40, and 60 iterations for the $balance - scale$ base example

growing for the leading edges (pairs of descriptors), and the connections that have been assigned low pheromone trail values gradually start to evaporate.

3.4.2 Pheromone Trail and Heuristic Function

Tables 3.4, 3.5 and Fig. 3.12 present the results of experiments dealing with the application of the pheromone trail and the heuristic function. These are meaningful findings because their analysis allows us to determine the role of pheromone trail in our final outcome. In order to achieve this goal, three different versions of our algorithm have been compared: classical—with the pheromone trail and the heuristic function together; with the pheromone trail only (used in the main transition rule); and with the heuristic function only.

As can be seen from Table 3.4, by far the best results in terms of classification accuracy can be obtained with the classical version of ACDT (using both the pheromone

Table 3.4 Comparison of results for three modes of algorithm operation (colony size—15 ant agents, 60 iterations)

Data set	Heuristic function and pheromone			Heuristic function			Pheromone		
	Accuracy	Num. of nodes	Height	Accuracy	Num. of nodes	Height	Accuracy	Num. of nodes	Height
australian	**0.8428**	15.5	6.3	0.8278	3.9	2.3	0.7913	37.5	9.5
balance-scale	0.7974	48.8	10.2	0.7821	43.3	8.8	**0.8102**	58.8	10.8
breast-cancer	**0.7258**	7.9	5.0	0.7007	6.4	4.3	0.6996	14.8	7.0
breast-tissue	**0.4921**	10.8	5.7	0.4667	8.8	5.0	0.4131	11.0	5.9
car	**0.9592**	45.0	10.6	0.9582	41.8	10.6	0.9046	65.2	11.4
cleveland	**0.5475**	19.3	6.9	0.5441	11.4	6.0	0.5416	14.2	6.2
CTG	**0.9908**	20.3	9.8	0.9850	19.0	10.3	0.6771	95.6	14.3
dermatology	**0.9530**	9.6	7.4	0.9374	8.8	7.4	0.8705	24.8	11.6
ecoli	0.8058	14.5	6.1	**0.8078**	12.0	5.9	0.7781	20.2	8.6
glass	**0.6675**	14.5	6.2	0.6452	12.6	5.8	0.5996	19.4	7.7
heart	**0.7748**	14.7	6.0	0.7744	9.3	5.0	0.7564	20.5	8.0
hepatitis	0.7948	5.8	4.4	0.7941	2.5	2.2	**0.7982**	9.9	6.0
horse-colic	0.8177	10.2	5.3	**0.8253**	5.9	4.3	0.8200	15.0	6.8
house-votes-84	**0.9342**	4.5	2.8	0.9298	2.7	2.0	0.9281	7.5	4.3
hungarian	**0.6426**	13.1	6.7	0.6402	8.7	6.0	0.6301	7.0	4.2
imports-85	**0.6870**	20.5	7.2	0.6653	17.4	6.8	0.5602	23.6	9.0
ionosphere	**0.8893**	8.1	5.3	0.8839	6.0	4.3	0.8556	15.2	7.7
iris	**0.9589**	3.5	3.2	0.9458	3.4	3.1	0.9252	6.8	5.4
jsbach	**0.7050**	288.6	13.5	0.6985	232.9	13.9	0.6375	260.0	16.0
kr-vs-kp	**0.9865**	23.0	10.5	0.9851	24.3	11.8	0.9105	98.9	15.7

(continued)

Table 3.4 (continued)

Data set	Heuristic function and pheromone			Heuristic function			Pheromone		
	Accuracy	Num. of nodes	Height	Accuracy	Num. of nodes	Height	Accuracy	Num. of nodes	Height
mushroom	0.6294	79.8	13.1	**0.6368**	50.1	11.6	0.6295	153.7	16.9
nursery	**0.9882**	160.5	13.5	0.9877	141.0	12.9	0.9314	347.5	15.2
optdigits	0.8084	162.9	11.7	**0.8270**	135.3	11.8	0.6609	307.0	16.4
parkinsons	**0.8782**	7.2	4.5	0.8517	5.9	4.0	0.8358	11.9	7.0
shuttle	**0.9972**	64.7	10.9	0.9971	56.9	10.8	0.9826	412.2	21.3
soybean	**0.8533**	26.8	9.1	0.8357	25.5	8.6	0.7252	38.5	11.4
transfusion	0.7133	6.8	2.8	**0.7134**	1.0	1.0	**0.7134**	2.5	1.5
wine	**0.8896**	6.6	4.3	0.8703	5.6	3.8	0.8419	12.3	7.3
yeast	0.5380	65.5	12.2	**0.5397**	41.7	10.5	0.4665	52.3	10.7
zoo	**0.8372**	6.4	4.8	0.8340	6.5	4.9	0.8081	8.0	5.6

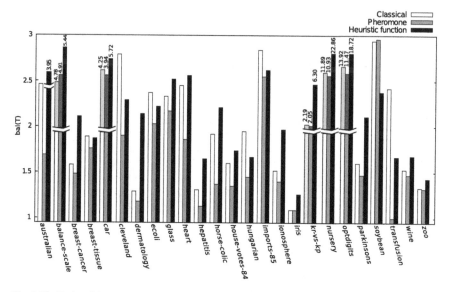

Fig. 3.12 Ratio of the number of nodes to the decision tree size—$bal(T)$ for most data sets

trail and the heuristic function). The classical algorithm produced better results than an algorithm that uses the heuristic function only, or the pheromone trail only, in as many as 16 cases. In six cases the results obtained with the classical algorithm are similar to those produced by an algorithm that uses the heuristic function only, whereas in the other eight cases the results are similar for each version of the algorithm, or slightly better when only the heuristic or only the pheromone was used (this is most often true for not very complex data sets).

Such observations were predictable, since ACDT is a learning algorithm based on additional knowledge. An algorithm based on the ACO approach improves the results via reinforcement learning. In consequence, the performance of ACDT is more affective compared to the version without this additional knowledge (see Sect. 3.4.1).

A very important matter is analysis of decision tree size in terms of the number of nodes as well as the tree height. Such an analysis is carried out using the decision tree balance ratio, determined as the ratio of the number of nodes to the decision tree height. More precisely:

$$bal(T) = \frac{n}{h}, \tag{3.22}$$

where n is the number of nodes (without terminal nodes) and h is the tree height, that is, the maximum length of a path from the root to a terminal node.

It can be easily observed in Fig. 3.12 that the most balanced decision trees (3.22) are obtained by using both the heuristic and the pheromone trail.

When an algorithm that is unable to convey information via the pheromone trail (and therefore to learn) is run, very degenerate decision trees are constructed. They

Fig. 3.13 Number of nodes
and decision tree height
depending on the version of
ACDT for the exemplary
australian base

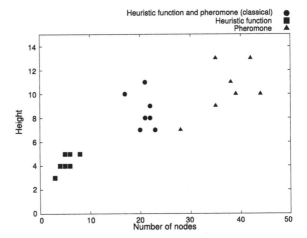

are trees with the number of nodes often close to the tree height, e.g. trees that are only regular on one side. As for the classification, this necessitates checking very long rules (i.e., paths in the tree). In the literature, there are many articles dealing with optimization of decision tree height, e.g. [12].

For an algorithm in which ant agents do not use external information (in the form of a heuristic function), decision trees are much more balanced. However, they also have a large number of nodes—which increases the memory volume used and the number of possible paths, i.e. rules, in a given tree. The classic combination of learning via both pheromone trails and external information in the form of a heuristic function enables construction of smaller decision trees than those constructed by using the pheromone trail only, with a much more balanced structure than that of trees generated by using the heuristic only. Therefore, we can say that the solution employed in ACDT (3.13) provides a balance between the two criteria—the number of nodes and the decision tree height

This can be observed especially in Fig. 3.13 (which is very important for this experiment) for the exemplary data set *australian*—and similar charts are created for other data sets. Here we can observe a correlation between the number of nodes and the tree height depending on the applied approach (for each version, we present several runs of the algorithm). In the case of using heuristic function only, each time the algorithm produces small decision trees, with few nodes, that are also degenerate.

In turn, the use of pheromone trail alone results in construction of large decision trees. However, combination of both elements in the main transition rule allows for obtaining balanced decision trees.

Figure 3.12 presents the results that confirm the justification for the classical approach that is typical of ACO, and also for constructing decision trees. We can notice that very often the value of the decision tree balance ratio (3.22) for an algorithm using the heuristic function only does not even exceed 1.5, whereas an algorithm using both the heuristic function and the pheromone trail enables construction of much more balanced trees—which are in addition also smaller than those constructed by an algorithm using the pheromone trail only.

Table 3.5 shows the results obtained with other measures for evaluating the classification quality for binary data sets (for more about these measures, see Chap. 4). White rows of the table present the results obtained with classifier quality evaluation measures for the classical algorithm (using both the heuristic function and the pheromone trail), whereas the gray rows present such results for an algorithm using the pheromone trail only. In most cases, the results confirm that, for each of the measures, the best results can be achieved by using the classical version of the algorithm. Only for the $horse - colic$ data set does each of the measures produce better results for an algorithm that uses solely the pheromone, which shows that there is little sense in using the heuristic in ACDT for data sets whose structure is similar to that of the $horse - colic$. The disparities between the results obtained for each of the measures are similar to the disparities in the classification accuracy.

In order to examine the relationship between the pheromone trail and the heuristic function, values of parameter β from the formula (3.13) other than the standard ones have been analyzed. Besides the standard value of $\beta = 3$, the values 1, 2 and 4 were also considered. In a further part of our work we also examined the lack of heuristic function (that is, $\beta = 0$, Table 3.4. The results presented in Table 3.6 confirm that the best results are mainly obtained for higher values of β—for most data sets, these are $\beta = 3$ or $\beta = 4$.

Table 3.5 Quality evaluation measures for binary data sets ($S(T)$—accuracy)

Data set	Accuracy	Recall	Precision	F1 score	MCC
Heuristic function and pheromone (classical)					
australian	**0.8428**	**0.8496**	**0.8121**	**0.8271**	**0.6889**
breast-cancer	**0.7258**	**0.3213**	**0.5621**	**0.3916**	**0.2855**
hepatitis	0.7948	0.8985	**0.8528**	0.8719	0.3949
horse-colic	0.8177	0.7168	0.8063	0.7495	0.6177
parkinsons	**0.8782**	0.9214	**0.9204**	**0.9189**	**0.6804**
Pheromone only					
australian	0.7913	0.7647	0.7713	0.7619	0.5837
breast-cancer	0.6996	0.2886	0.5436	0.3396	0.2536
hepatitis	**0.7982**	**0.9221**	0.8426	**0.8767**	**0.4042**
horse-colic	**0.8200**	**0.7234**	**0.8089**	**0.7532**	**0.6243**
parkinsons	0.8358	**0.9359**	0.8727	0.9013	0.5629

3.4.3 Comparison with Other Algorithms

To compare the performance of different algorithms, we consider two evaluation criteria for decision trees: accuracy and tree size, for the analyzed approaches. We have compared the ACDT algorithm with the random trees classifier (implemented in the WEKA system and described in [6]), C4.5, CART (both also implemented in

Table 3.6 Comparative study for different values of the β parameter (with $\alpha = 1$)

Dataset	$\beta = 1$	$\beta = 2$	$\beta = 3$	$\beta = 4$
australian	0.8403	0.8412	0.8428	0.8499
balance-scale	0.8020	0.7958	0.7974	0.7937
breast-cancer	0.7196	0.7168	0.7258	0.7207
breast-tissue	0.4820	0.4737	0.4921	0.4923
car	0.9552	0.9561	0.9592	0.9588
cleveland	0.5382	0.5398	0.5475	0.5442
CTG	0.9703	0.9857	0.9908	0.9927
dermatology	0.9541	0.9521	0.9530	0.9583
ecoli	0.8022	0.8095	0.8058	0.8144
glass	0.6693	0.6569	0.6675	0.6497
heart	0.7716	0.7785	0.7748	0.7796
hepatitis	0.7956	0.7965	0.7948	0.7920
horse-colic	0.8128	0.8152	0.8177	0.8167
house-votes-84	0.9323	0.9302	0.9342	0.9325
hungarian	0.6422	0.6378	0.6426	0.6462
imports-85	0.6521	0.6708	0.6870	0.6915
ionosphere	0.8862	0.8898	0.8893	0.8871
iris	0.9550	0.9591	0.9589	0.9591
jsbach	0.6807	0.6968	0.7050	0.7100
kr-vs-kp	0.9822	0.9856	0.9865	0.9867
mushroom	0.6384	0.6331	0.6294	0.6399
nursery	0.9852	0.9877	0.9882	0.9885
optdigits	0.7833	0.7981	0.8084	0.8184
parkinsons	0.8784	0.8768	0.8782	0.8678
shuttle	0.9968	0.9972	0.9972	0.9971
soybean	0.8348	0.8515	0.8533	0.8541
transfusion	0.7131	0.7134	0.7133	0.7136
wine	0.8878	0.8840	0.8896	0.8861
yeast	0.5176	0.5366	0.5380	0.5448
zoo	0.8146	0.8381	0.8372	0.8379

Table 3.7 Comparison of ACDT with other algorithms

Data set	ACDT		Random trees		CART		Ant-Tree-Miner		C4.5	
	Accuracy rate	No. of nodes	Accuracy rate	No. of nodes	Accuracy rate	No. of nodes	Accuracy rate	No. of nodes	Accuracy rate	No. of nodes
australian	0.8428	15.5	0.7266	363.0	0.8391	**4.6**	0.8416	47.6	**0.8478**	28.6
balance-scale	**0.7974**	48.8	0.6983	507.5	0.7935	49.8	0.6299	**42.8**	0.6367	43.0
breast-cancer	0.7258	7.9	0.6174	462.0	0.7175	**5.8**	**0.7332**	12.5	0.7197	10.8
breast-tissue	0.4921	**10.8**	0.4555	47.6	0.6376	14.6	**0.6451**	22.2	0.5871	24.8
car	**0.9592**	**45.0**	0.8567	255.4	0.9184	55.0	0.8693	64.2	0.8785	73.0
cleveland	**0.5475**	19.3	0.4844	237.1	0.5408	**16.6**	0.5314	60.1	0.5087	76.7
CTG	0.9908	20.3	0.8903	152.3	0.9979	19.0	**1.0000**	**19.0**	**1.0000**	**19.0**
dermatology	**0.9530**	**9.6**	0.8646	140.7	0.9373	14.0	0.9444	25.9	0.9320	36.1
ecoli	0.8058	**14.5**	0.7660	113.6	0.8012	19.2	0.8188	31.9	**0.8252**	34.8
glass	0.6675	**14.5**	0.6500	92.4	0.6802	28.4	**0.6901**	40.0	0.6803	46.6
heart	**0.7748**	**14.7**	0.7138	126.0	0.7703	17.8	0.7644	30.6	0.7590	33.5
hepatitis	0.7948	**5.8**	0.7738	87.4	0.7817	5.9	**0.8075**	12.8	0.7933	18.4
horse-colic	0.8177	10.2	0.7757	469.3	0.8253	**7.0**	0.8250	14.8	**0.8386**	7.7
house-votes-84	0.9342	**4.5**	0.9026	113.1	0.9312	9.6	0.9331	10.8	**0.9493**	9.8
hungarian	0.6426	13.1	0.5816	220.0	0.6279	**5.2**	0.6478	36.6	**0.6696**	51.4
imports-85	0.6870	**20.5**	0.6757	207.3	**0.8088**	45.6	0.7541	53.4	0.7612	63.2
ionosphere	0.8893	**8.1**	0.8433	50.6	0.8857	16.2	0.8872	21.5	**0.8991**	25.8

(continued)

Table 3.7 (continued)

Data set	ACDT		Random trees		CART		Ant-Tree-Miner		C4.5	
	Accuracy rate	No. of nodes	Accuracy rate	No. of nodes	Accuracy rate	No. of nodes	Accuracy rate	No. of nodes	Accuracy rate	No. of nodes
iris	**0.9589**	**3.5**	0.9395	19.4	0.9392	8.2	0.9542	7.6	0.9390	8.0
jsbach	0.7050	**288.6**	0.6678	3126.3	**0.7423**	431.0	0.7231	415.5	0.7406	740.0
kr-vs-kp	0.9865	**23.0**	0.9409	667.0	0.9803	59.0	0.9720	51.8	**0.9906**	45.0
mushroom	0.6294	79.8	0.5004	3196.7	**0.6558**	**33.0**	0.5966	151.7	0.6081	128.0
nursery	**0.9882**	**160.5**	0.9440	1147.3	0.9775	303.0	0.9744	186.5	0.9698	368.0
optdigits	0.8084	162.9	0.7674	517.0	0.7957	**113.0**	0.8051	219.8	**0.8189**	191.0
parkinsons	0.8782	**7.2**	0.8718	34.2	0.8716	10.3	**0.9011**	14.3	0.8824	19.6
shuttle	**0.9972**	**64.7**	0.9960	275.5	0.9968	103.0	0.9966	74.4	0.9971	109.0
soybean	0.8533	**26.8**	0.7751	387.6	**0.8729**	70.0	0.8687	69.9	0.8463	83.1
transfusion	0.7133	**6.8**	0.6357	370.6	**0.7411**	11.8	0.7176	23.3	0.7371	12.2
wine	0.8896	**6.6**	0.8663	22.8	0.9001	9.4	**0.9476**	9.8	0.9441	9.2
yeast	**0.5380**	**65.5**	0.4878	366.0	0.5345	73.0	0.5163	126.8	0.5112	150.0
zoo	0.8372	**6.4**	0.7857	35.0	0.8557	14.2	**0.8977**	16.9	0.8757	18.6

Table 3.8 Friedman Test results and mean ranks (the best rank in boldface)—four modes of algorithm operation

	Values
N	30
Chi-Square	71.8294
Degrees of freedom	3
p value less than	0.0001
5% critical difference	0.3023
Mean ranks	
ACDT (std.)	**1.3667**
Pheromone only	2.7500
Heuristic only	1.8833
Without h. and ph.	4.0000

WEKA as j48 and simpleCart, respectively), and Ant-Tree-Miner in [8, 9] (Table 3.7).

All of our experimental results have confirmed that in terms of accuracy ACDT is similar to or better than the tested classical approaches and other algorithms based on ACO. However, in terms of decision tree size (number of nodes), ACDT is significantly better than other algorithms (see Tables 3.7, 3.10 and 3.11).

Taking into account both aspects, the experimental results confirm that application of ant colony optimization algorithms to construction of classifiers allows for better analysis of the entire solution space, which results in improved object classification.

Comparison with other algorithms, carried out with help of preprocessing (as described at the beginning of this chapter), confirms that in this setting ACDT is significantly better than other algorithms—as shown in Table 3.12 and described in Sect. 3.4.4 (statistical analysis).

3.4.4 Statistical Analysis

Experimental results of ACDT methods are compared using a non-parametric statistical hypothesis test, i.e., the Friedman test for $\alpha = 0.05$. Parameters of the Friedman test are shown in Tables 3.8 and 3.10. In case of Table 3.10, the accuracy and the number of nodes were analyzed. The 5% critical difference and the p-values should be carefully considered. Additionally, the average rank values for the analyzed approaches are also recorded (bold font indicates the best result). Table 3.9 presents the average rank values for 30 algorithm runs: the lower the rank, the better the algorithm result. The chi-square values and the p-values for every analysis are also presented. Although Table 3.11 presents the rank differences between the compared algorithms, note that values below zero indicate that the analyzed algorithm

Table 3.9 Collection of mean ranks for the Friedman Test (for $df = 3$)

Data set	ACDT (std.)	Onlypheromone	Only heuristic	Without h.and ph.	χ^2	$p \leq$
australian	**1.6**	2.7	1.8	4.0	66.25	0.0001
balance-scale	2.0	**1.5**	2.8	3.7	48.81	0.0001
breast-cancer	**1.9**	2.7	2.7	2.8	9.37	0.0248
breast-tissue	**1.6**	2.7	2.0	3.8	48.87	0.0001
car	**1.2**	3.5	1.8	3.5	74.93	0.0001
cleveland	**2.0**	2.6	2.5	2.9	7.40	0.0602
CTG	**1.3**	3.0	1.7	4.0	32.80	0.0001
dermatology	**1.3**	2.9	1.8	4.0	77.56	0.0001
ecoli	1.8	2.5	**1.7**	4.0	61.41	0.0001
glass	**1.6**	2.6	1.8	4.0	63.49	0.0001
heart	**1.7**	1.8	2.5	4.0	61.41	0.0001
hepatitis	2.4	**2.2**	2.6	2.7	2.63	0.4523
horse-colic	2.4	2.2	**1.5**	4.0	61.93	0.0001
house-votes-84	**1.7**	2.4	1.9	4.0	58.48	0.0001
hungarian	**2.3**	2.6	**2.3**	2.7	2.52	0.4714
imports-85	**1.3**	2.9	1.8	4.0	77.56	0.0001
ionosphere	**1.5**	2.8	1.8	4.0	59.54	0.0001
iris	**1.5**	2.7	1.9	4.0	64.21	0.0001
jsbach	**1.3**	3.0	1.7	4.0	32.80	0.0001
kr-vs-kp	**1.2**	3.6	1.8	3.4	74.72	0.0001
mushroom	**2.2**	2.3	1.5	4.0	18.3	0.0004
nursery	**1.3**	3.1	1.7	3.9	78.76	0.0001
optdigits	1.7	3.1	**1.4**	3.9	79.57	0.0001
parkinsons	**1.6**	2.4	2.1	3.9	54.76	0.0001
shuttle	**1.3**	3.0	1.7	4.0	25.23	0.0001
soybean	**1.4**	2.9	1.7	4.0	78.16	0.0001
transfusion	2.4	2.4	**1.8**	3.5	30.07	0.0001
wine	**1.6**	2.5	1.9	4.0	61.24	0.0001
yeast	**1.6**	2.9	1.6	3.9	27.80	0.0001
zoo	**2.0**	2.5	2.2	3.4	22.29	0.0001

(row in the table) is worse than the algorithm it is compared with (column in the table). Critical differences are marked in bold font.

Tables 3.8 and 3.9 present the Friedman test results for each analyzed data set for 30 representations. The statistics confirm that the classical version of ACDT yields the best results, thanks to its balanced treatment of the pheromone trail and the heuristic function,. The smaller the mean ranks for the sample values, the better

Table 3.10 Friedman Test results and mean ranks (the best rank in boldface)

Accuracy		Number of nodes	
	Values		Values
N	30	N	30
Chi-Square	47.6194	Chi-Square	86.3154
Degrees of freedom	4	Degrees of freedom	4
p value less than	0.0001	p value less than	0.0001
5% critical difference	0.6315	5% critical difference	0.4102
Mean ranks			
ACDT	**2.3667**	ACDT	**1.4333**
Random trees	4.7667	Random trees	5.0000
CART	2.7000	CART	2.1333
Ant-Tree-Miner	2.6167	Ant-Tree-Miner	3.0000
C4.5	2.5500	C4.5	3.4333

Table 3.11 Differences between algorithms (critical differences in boldface)

	ACDT	Random trees	CART	Ant-Tree-Miner	C4.5
Accuracy					
ACDT	–	−2.4000	−0.3333	−0.2500	−0.1833
Rand. t.	**2.4000**	–	**2.0667**	**2.1500**	**2.2167**
CART	0.3333	−2.0667	–	0.0833	0.1500
A-T-M	0.2500	−2.1500	−0.0833	–	0.0666
C4.5	0.1833	−2.2167	−0.1500	−0.0667	–
Number of nodes					
ACDT	–	−3.5667	−0.7000	−1.5667	−2.0000
Rand. t.	**3.5667**	–	**2.8667**	**2.0000**	**1.5667**
CART	**0.7000**	−2.8667	–	−0.8667	−1.3000
A-T-M	**1.5667**	−2.0000	**0.8667**	–	−0.4333
C4.5	**2.0000**	−1.5667	**1.3000**	0.4333	–

the given approach (the best values are indicated by boldface font). Table 3.9 presents the results for degrees of freedom (df = 3).

The Friedman test confirms that ACDT is significantly better than random trees, and that no significant differences (in terms of accuracy) can be observed compared to other approaches. What is more, in terms of decision tree size, the ACDT algorithm is also significantly better than other algorithms. Table 3.11 presents a statistical comparison of ACDT and other discussed approaches. Bold fonts indicate the values that fulfill the criterion of 5% critical difference (the smaller the value, the better the

Table 3.12 Statistical results of comparison with other algorithms (better, no difference, worse)

	ACDT	Random trees	CART	Ant-Tree-Miner	C4.5
ACDT	–	(30, 0, 0)	(21, 4, 5)	(18, 4, 8)	(22, 1, 7)
Rand. t.	(0, 0, 30)	–	(3, 1, 26)	(2, 0, 28)	(4, 0, 26)
CART	(5, 4, 21)	(26, 1, 3)	–	(13, 2, 15)	(15, 3, 12)
A-T-M	(8, 4, 18)	(28, 0, 2)	(15, 2, 13)	–	(15, 7, 8)
C4.5	(7, 1, 22)	(26, 0, 4)	(12, 3, 15)	(8, 7, 15)	–

Table 3.13 Statistical comparison results (better, no difference, worse)—data sets subjected to preprocessing

	ACDT (std.)	Only pheromone	Only heuristic	Without heuristic and pheromone
ACDT (std.)	–	(25, 2, 3)	(20, 5, 5)	(30, 0, 0)
Pheromone	(3, 2, 25)	–	(2, 3, 25)	(30, 0, 0)
Heuristic	(5, 5, 20)	(25, 3, 2)	–	(30, 0, 0)
Without h. and ph.	(0, 0, 30)	(0, 0, 30)	(0, 0, 30)	–

algorithm, statistically speaking). Table 3.10 presents all values determined during the statistical analyses, including the mean ranks for samples.

The Wilcoxon Signed-Rank test for $\alpha = 0.05$ also confirms that ACDT is better than other approaches. The conducted Wilcoxon Signed-Rank tests confirm the results obtained when e.g. comparing the classification accuracy or the average value of $n_{s/r} = 30$, and that the analyzed W parameter is most often equal to 250–300.

Though ACDT is better than the average ranks obtained by all other algorithms, we also wanted to compare different versions of ACDT to check the algorithm operation. The most important statistical analyses are presented in Tables 3.12 and 3.13. The statistical analysis confirms the observations as described above.

3.5 Conclusions

The good results produced by ACO algorithms (when used to solve "graph problems") can also be observed for ACDT, as has been confirmed by the results of experiments. A comparison with other algorithms shows that ACDT is capable of achieving similar results. Additionally, the height of classification trees is reduced. A comparison in terms of other quality evaluation measures (see Table 3.5) shows that high classification quality is also achieved for these measures.

We justify the need for using both the heuristic function and the pheromone trail if balanced results are desired. The results confirm that ant agents learn via the pheromone trail. Therefore, we can say that ACDT enables optimization of decision tree construction based on the heuristic function. If we treated as a local optimum a single decision tree constructed with an algorithm whose heuristic function was used for that construction, then ACDT would first explore, and then exploit, the space near this solution.

In case of medium-sized or small data sets (in terms of the numbers of attributes and attribute values), pheromone maps clearly present a data set as the solution space (representing the hypothesis space), where the pheromone trail values laid on the edges represent potentially good solutions. Change in the pheromone trail value (which can be seen in the charts presented here) and gradual improvement in the quality of constructed decision trees show that learning via the pheromone trail yields good results. Analysis of pheromone maps at different stages of algorithm operation shows that ACDT searches various subspaces of the solution space. This characteristic cannot be obtained using classical approaches, and is similar to what should be achieved for sets of classifiers. Statistical evidence confirms these results.

The obtained results, described in Sect. 3.4, can be treated as complement to the approach defined in this chapter. Statistical analysis of the conducted experiments allow us to claim that application of the ant colony optimization algorithm to the decision tree learning process allows us to find new, alternative, and often better solutions (decision trees). Reinforcement learning with the pheromone trail allows for finding locally better solutions, which by further improvement enable adaptation of the learned decision tree to the analyzed problem, even if at the initial phase those solutions are not satisfactory.

A good confirmation of the algorithm's efficiency, and at the same time explanation how the adaptation works, is provided by the pheromone maps. Pheromone maps allow us to track an indirect algorithm solution and present its work.

The pheromone map is of secondary importance in the final result yielded by the generated classifier, though it corresponds to the structure of the generated decision tree. Yet in the analysis of algorithm operation it allows for justifying why the algorithm is effective with the combination of the heuristic (classical solution), stochastics, and learning with feedback through pheromone trail.

Other observations regarding ACDT should also be mentioned. Boryczka and Kozak [4] analyzed the impact of changing the decision tree size measure on the achieved results. In order to compare the impact of the method for evaluating the growth (and indirectly—quality) of the decision tree constructed by ant agents, we have proposed three new measures.

All of the approaches differ from the classical version in the formula for calculating decision tree growth (3.11), whereas (3.10) remains unchanged. Accordingly, the following options are presented:

- decision tree growth is determined by the multiplicative inverse of its height

$$s_2(T) = \frac{1}{h}, \tag{3.23}$$

where h is the tree height.

- decision tree growth is determined by the multiplicative inverse of the sum of its height and the number of nodes (n)

$$s_3(T) = \frac{1}{h + n},$$ (3.24)

- decision tree growth is determined by the multiplicative inverse of the product of its height and the number of nodes

$$s_4(T) = \frac{1}{h \cdot n}.$$ (3.25)

Obviously, in each of these approaches, $s(T)$ should be replaced in Eq. (3.10) with $s_2(T)$, $s_3(T)$ or $s_4(T)$, respectively.

In [4] the authors noted that the evaluation method has significant impact on the results of ACDT. For the majority of testable data bases, the applied evaluation of decision tree growth causes higher resistance to changes in parameter ϕ (3.24). This reduces the impact of decision tree growth on the evaluation of decision tree quality. As ant agents construct a decision tree based on the accuracy of classification (3.12), the resulting trees are larger. A similar relationship can be observed for $s_4(T)$ (3.25). In the same article, the authors analyzed the Pareto front to consider the effectiveness of the ϕ and ψ values (3.10). The described observations confirm that in case of quality evaluation in the context of ant colony optimization algorithms the choice of the best decision tree is made based on two criteria, according to Eq. (3.10).

References

1. A.G. Barto, *Reinforcement learning: An Introduction*, (MIT Press, 1998)
2. U. Boryczka, J. Kozak, Ant colony decision trees—a new method for constructing decision trees based on ant colony optimization, in *Computational Collective Intelligence, Technologies and Applications*, ed. by J.-S. Pan, S.-M. Chen, N. Nguyen. Lecture Notes in Computer Science, vol. 6421 (Springer, Berlin, Heidelberg, 2010), pp. 373–382
3. U. Boryczka, J. Kozak, New insights of cooperation among ants in ant colony decision trees, in *Third World Congress on Nature and Biologically Inspired Computing, NaBIC*, Salamanca, Spain, pp. 255–260, 19-21 Oct 2011
4. U. Boryczka, J. Kozak, Enhancing the effectiveness of ant colony decision tree algorithms by co-learning. Appl. Soft Comput. **30**, 166–178 (2015)
5. U. Boryczka, B. Probierz, J. Kozak, An ant colony optimization algorithm for an automatic categorization of emails, in *6th International Conference on Computational Collective Intelligence, Technologies and Applications, ICCCI* Seoul, Korea, pp. 583–592, 24–26 Sept 2014
6. R.R. Bouckaert, E. Frank, M. Hall, R. Kirkby, P. Reutemann, A. Seewald, D. Scuse, Weka manual for version 3-7-10 (2013)
7. L. Breiman, J.H. Friedman, R.A. Olshen, C.J. Stone, *Classification and Regression Trees* (Chapman & Hall, New York, 1984)
8. F.E.B. Otero, MYRA: an ACO framework for classification (2015), https://github.com/febo/myra

9. F.E.B. Otero, A.A. Freitas, C.G. Johnson, Inducing decision trees with an ant colony optimization algorithm. Appl. Soft Comput. **12**(11), 3615–3626 (2012)
10. J.R. Quinlan, *C4.5: Programs for Machine Learning*, (Morgan Kaufmann, 1993)
11. L. Rokach, O. Maimon. *Data Mining With Decision Trees: Theory And Applications*, (World Scientific Publishing, 2008)
12. B. Zielosko, I. Chikalov, M. Moshkov, T. Amin, Optimization of decision rules based on dynamic programming approach, in *Innovations in Intelligent Machines*, vol. 4, (Springer, 2014), pp. 369–392

Chapter 4
Adaptive Goal Function of the ACDT Algorithm

The above approach entails an adaptive goal function of ACDT, which is aimed at evaluating classification based on various quality measures. ACDT with the adaptive goal function arises as a result of changing the goal function (solution quality evaluation) of an ant agent while maintaining the method of determining the value of the heuristic function. This gives a significant advantage to ACO based learning over other, classical approaches, which do not allow for adapting a classifier to a given measure while the node division criteria remain invariant. Consequently, one can construct trees with better recall, precision, F-measure or the Matthews correlation coefficient, etc., depending on the current needs by using a different classification evaluation as the goal function of ant colony optimization algorithms.

4.1 Evaluation of Classification

To make the contents of this chapter clearer to the reader, in this section we explain in detail the problems involved in classification quality evaluation, and define example measures which were used in the experiments. Classification quality evaluation is one of basic machine learning problems, for it underlies the judgment if a given classifier can be judged as good or bad. However, there are no classifiers that can be modified based on the measure used, or optimized according to several measures. Yet in real-world problems it is often more important, for example, to ensure precision or balance of two different classification quality evaluation measures [1].

In this chapter we present selected measures of binary classification quality (for sets with two decision classes) which can be estimated based on a confusion matrix (Table 4.1). Thanks to the confusion matrix, it is possible to better evaluate the classification quality based on the information on the actual decision class of the object and the class that the object has been classified to [3]. Such information allows for

© Springer International Publishing AG, part of Springer Nature 2019
J. Kozak, *Decision Tree and Ensemble Learning Based on Ant
Colony Optimization*, Studies in Computational Intelligence 781,
https://doi.org/10.1007/978-3-319-93752-6_4

Table 4.1 Confusion matrix

	Predicted positive	Predicted negative
Positive examples (P)	True positive (TP)	False negative (FN)
Negative examples (N)	False positive (FP)	True negative (TN)

estimating the values of accuracy, recall, precision, F-measure and Matthews correlation coefficient, each time represented by the classification evaluation $ev(T, S)$ for decision tree T and data set S, whereby the appropriate abbreviation is used for each measure.

Table 4.1 presents the form of a confusion matrix. To simplify further interpretation, symbols P and N are used for the two classes under consideration.

It should be noted that classification accuracy is a measure that only shows how many objects have been classified correctly. As opposed to this, precision allows for assessing the confidence with which one can assume that an object actually belonging to a given class are classified correctly. This is particularly important when the classification priority is that all objects actually belonging to a given class are correctly classified to that class. In such a case, it is better to assign too many objects to the given class than to omit one of the relevant objects. In turn, in case of recall, the information of paramount importance is whether a given class only contains objects that have been correctly assigned to it. Accordingly, it is better to omit one of the relevant objects than to incorrectly assign an irrelevant object to that class.

Accuracy is one of the most popular measures of classification evaluation. It should, however, be noted that this measure does not provide sufficient evaluation, for example, for data sets with a considerable diversity of decision classes. This was observed, among others, in [1]. The accuracy of classification describes the ratio of objects that have been classified correctly to all objects in a given class.

$$ev_{acc}(T, S) = \frac{(TP + TN)}{(TP + TN + FP + FN)}. \tag{4.1}$$

Recall is a different, simple measure used for binary classification, whose value is given by Eq. (4.2). Accordingly, recall is the ratio of objects that have been correctly classified to class P to all objects that should have been classified to this class. Therefore, in case of recall, it is better to incorrectly assign an object belonging to class N to class P than to incorrectly classify an object belonging to class P.

$$ev_{rec}(T, S) = \frac{TP}{(TP + FN)}. \tag{4.2}$$

Precision is a measure that evaluates a classifier based on incorrect classification of objects belonging to class N to class P. Here it is better to omit certain objects from class P than to incorrectly assign to class P some objects belonging to class N.

Precision is determined based on the ratio of objects that have been correctly classified to class P to all objects that have been assigned to that class:

$$ev_{prec}(T, S) = \frac{TP}{(TP + FP)}. \tag{4.3}$$

F-measure (F1 score) constitutes an attempt to balance precision and recall. This measure is frequently used to evaluate binary classification (two decision classes). In its simplest form, it is determined as the ratio of the product of precision and recall to the sum of these two measures:

$$ev_{fm}(T, S) = 2 \cdot \frac{ev_{prec}(T, S) \cdot ev_{rec}(T, S)}{ev_{prec}(T, S) + ev_{rec}(T, S)}, \tag{4.4}$$

After the appropriate substitutions, the following formula is obtained:

$$ev_{fm}(T, S) = \frac{(TP + TP)}{(TP + TP + FP + FN)}. \tag{4.5}$$

Matthews correlation coefficient is also used as an evaluation measure for binary classification. In contrast to the above-mentioned recall, precision and F-measure, this coefficient is determined based on the entire confusion matrix, and calculated using the following formula:

$$ev_{mcc}(T, S) = \frac{(TP \cdot TN - FP \cdot FN)}{\sqrt{(TP + FP) \cdot (TP + FN) \cdot (TN + FP) \cdot (TN + FN)}}. \tag{4.6}$$

The Matthews correlation coefficient is considered a good measure to be used in case of very large disproportion between the cardinalities of decision classes.

4.2 The Idea of Adaptive Goal Function

This approach to the ACDT algorithm is based on changing the goal function of ant agents by modifying the way of determining the evaluation function for estimating decision tree quality (3.10)—which, for example, influences pheromone trail updates (3.18). This will result in rewarding solutions with a higher value of a given classification evaluation measure with the pheromone trail.

It should be kept in mind that the heuristic function is not changed in this approach (however, there are still such possibilities). Thus, the most often created decision trees are those adapted to achieving a better quality of classification according to the given split criterion. When other classification measure is used for evaluating the solution quality, the decision trees are created in some sense on the basis of several criteria. In consequence, decision trees created in the classical way are optimized with respect

to the selected measure. In this case, evaluation of the ant agent quality given by the
formula (3.10) is defined as follows:

$$Q(T) = \phi \cdot w(T) + \psi \cdot ev(T, S), \tag{4.7}$$

where $ev(T, S)$ denotes the selected method of evaluating the classification generated by decision tree T constructed using an ant agent based on data set S. The value
of $ev(T, S)$ is determined depending on the selected measure according to the following formulas: accuracy (4.1), recall (4.2), precision (4.3), f-measure (4.5), and
the Matthews correlation coefficient (4.6).

4.3   Results of Experiments

The purpose of experiments was to compare five different versions of ACDT with the
proposed classification quality evaluation measures, and thus to estimate the impact
of the measure used on the final results achieved. The way of evaluating classification
quality has a significant impact during the process of selecting the ant agent. During
that process, the pheromone trail is updated, so it has impact on the new value of the
updated pheromone. The values of each of the considered measures were taken into
account (independently of the actual method for estimating $ev(T, S)$). This means
that each classifier built was evaluated under the 5 analyzed classification quality
measures. The decision tree size was analyzed as well, along with the number of
nodes, the tree height and the algorithm uptime—classifier creation time.

To estimate the quality of proposed solutions, a series of experiments were conducted. Their results can be found in Table 4.3, Figs. 4.1 and 4.2, where the best
results are presented in bold. Every experiment was repeated 30 times, each time
for 200 generations with 15 ant agents. The remaining parameters were as follows:
$q_0 = 0, 3$, $\alpha = 3, 0$, $\gamma = 0, 1$, $\phi = 0, 05$ and $\psi = 1, 0$.

The evaluation of ACDT performance was carried out using 6 public domain data
sets from the UCI (University of California at Irvine) data set repository, selected
in such a manner that only two decision classes were available. Thus the analyzed
data sets were not the same as in Chap. 3. They were estimated by 10-fold cross-
validation. The characteristics of data sets which were important for carrying out the
experiments are as presented in Table 4.2.

Roughly speaking, the ratio of objects in class "0" to all objects is similar for the
australian and *heart* sets (44%), and for the *ttt* and *bcw* sets (65–66%). The number
of objects in class "0" compared to the data set size is the smallest for the *horse-colic*
set (37%) and the largest for the *hepatitis* set (79%).

The obtained results (presented in Table 4.3) confirm that in most cases the maximum value for each of the measures is achieved during the operation of the algorithm
using the given measure as its goal function. This is true irrespective of which data
set is analyzed. In other words, the highest accuracy is obtained when measure (4.1),
i.e. accuracy, is used in Eq. (3.10) as $ev(T, S)$, and the highest precision is obtained

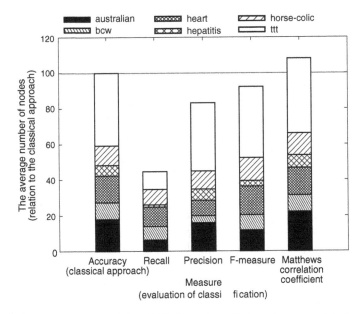

Fig. 4.1 Average fig/num of nodes established based on all data sets compared to the classical approach

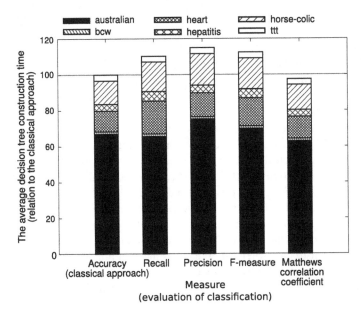

Fig. 4.2 Average decision tree construction time established based on all data sets compared to the classical approach

Table 4.2 Characteristics of data sets

Data set	No. of instances	No. of att.	No. of instances in class "0"	No. of instances in class "1"
australian	690	14	307	383
bcw	699	9	458	241
heart	270	13	120	150
hepatitis	155	19	123	32
horse-colic	366	22	136	230
ttt	958	9	625	332

when it is precision (4.3) that is optimized, etc. This means that we can maximize the value of a given measure by changing the way of determining $ev(T, S)$. What is interesting, this can often be achieved without considerable deterioration in accuracy (which is treated here as the measure originally used in the ACDT algorithm).

When comparing the results for the *australian* set, one can notice that accuracy remains at a similar level while recall improves by nearly 10% compared to the classical approach, and by not less than 5% when other measures are used. The situation for precision is similar (a 6% improvement). The disproportion is slightly smaller when F-measure is used. Very similar dependencies were found for the *hepatitis* set. However, differences in the classification accuracy were larger, and the highest value of accuracy was obtained when F-measure was used. These results are similar to those obtained for the *heart* set, where, additionally, recall made the classification accuracy considerably worse. Larger differences in accuracy, but with the previously described observations still holding true, can be noticed for the remaining data sets: *bcw*, *horse-colic* and *ttt*, i.e. those for which the difference in the number of objects belonging to decision classes is 33–35%.

F-measure can be regarded as the most universal goal function; when F-measure is used, the results are usually correct for each of the measures. This probably results from the way in which the values of F-measure are determined, as described in Sect. 4.1. F-measure somewhat balances precision and recall. The size of the trees constructed by using F-measure (Fig. 4.1) and the run time of the algorithm (Fig. 4.2) do not differ much either from the average values obtained in other cases.

Figure 4.1 presents a comparison between the number of nodes in the constructed trees with respect to classical ACDT (with accuracy). The smallest decision trees are definitely constructed when using recall as the goal function. Interestingly, even a large reduction in the number of nodes in the tree does not necessarily result in a large deterioration in accuracy, and often increases the value of recall.

Figure 4.2 presents a comparison between decision tree construction times. The algorithm with accuracy and MCC as the goal function is the fastest one, and slows down slightly for recall, F-measure and precision. One can, however, state that the algorithm works in a similar way for each of the measures.

Table 4.3 Test measure values by the goal function—average for all runs

Measure		australian					bcw				
		acc	rec	prec	F1	MCC	acc	rec	prec	F1	MCC
Goal function	acc	**0.846**	0.849	0.816	0.828	0.691	**0.937**	0.932	0.973	0.950	0.861
	rec	0.835	**0.932**	0.761	0.834	**0.692**	0.909	**0.956**	0.914	0.934	0.793
	prec	0.838	0.744	**0.878**	0.801	0.677	0.918	0.897	**0.979**	0.935	0.832
	F1	0.843	0.882	0.803	**0.838**	0.702	0.932	0.929	0.969	**0.952**	0.854
	MCC	0.833	0.838	0.804	0.817	0.670	0.935	0.932	0.973	0.949	**0.863**

Measure		heart					hepatitis				
		acc	rec	prec	F1	MCC	acc	rec	prec	F1	MCC
Goal function	acc	0.767	0.717	0.762	0.728	0.537	0.789	0.891	0.852	0.868	0.390
	rec	0.707	**0.798**	0.654	0.710	0.443	0.788	**0.966**	0.806	0.878	0.318
	prec	0.751	0.563	**0.849**	0.657	0.514	0.775	0.839	**0.874**	0.852	0.403
	F1	**0.780**	0.753	0.766	**0.749**	0.561	**0.814**	0.941	0.845	**0.888**	**0.410**
	MCC	0.776	0.727	0.777	0.736	**0.567**	0.789	0.889	0.858	0.867	0.404

Measure		horse-colic					ttt				
		acc	rec	prec	F1	MCC	acc	rec	prec	F1	MCC
Goal function	acc	**0.818**	0.733	0.798	**0.755**	**0.620**	0.858	0.925	0.868	0.895	0.681
	rec	0.764	**0.820**	0.667	0.729	0.543	0.761	**0.992**	0.736	0.845	0.465
	prec	0.815	0.637	**0.854**	0.720	0.608	0.823	0.837	**0.888**	0.859	0.627
	F1	0.809	0.726	0.786	0.745	0.603	0.858	0.922	0.869	**0.899**	0.679
	MCC	0.814	0.723	0.791	0.747	0.609	**0.863**	0.924	0.875	0.894	**0.693**

Abbreviations: *acc* accuracy; *rec* recall; *prec* precision; *F1* F-measure; *MCC* Matthews correlation coefficient

To unequivocally confirm results of the experiments and to enable achievement of their purpose, statistical tests were conducted as well. They involved a non-parametric statistical hypothesis test, i.e., the Friedman test for $\alpha = 0.05$. Results for the Friedman test are presented in Table 4.4, while average ranges achieved with these statistical tests are shown in Table 4.5. As may bee seen, the results in Table 4.4 are necessary for a proper analysis of the ranks (Table 4.5). In this case, ranks for every measure are given in columns. Thus the column for accuracy presents the ranks obtained after analysis of classification accuracy for ACDT using the proper classification quality evaluation.

Statistical results undoubtedly confirm the observations described in this chapter. We can see that each time the highest rank achieved is obtained for the algorithm with the same quality measure as that for which the results are analyzed. In case of recall and precision, the comparison with the remaining versions of the algorithm exceed the critical difference. Thus the algorithm with the precision measure is significantly better for the needs of achieving better precision, and analogically in case of recall.

Interestingly (though this could be predicted, knowing the measures), if the measure selected in the algorithm is precision, the results for the recall are significantly worse (always worse than for precision), and vice versa.

Table 4.4 Friedman test results—adaptive goal function of ACDT

Accuracy		Recall	
	Values		Values
N	5	N	5
Chi-square	12.9831	Chi-square	20.0000
Degrees of freedom	4	Degrees of freedom	4
p value is less than	0.0115	p value is less than	0.0006
5% critical difference	1.4015	5% critical difference	0.8481
Precision		F-measure	
	Values		Values
N	5	N	5
Chi-square	21.2773	Chi-square	15.4667
Degrees of freedom	4	Degrees of freedom	4
p value less than	0.0004	p value less than	0.0039
5% critical difference	0.6997	5% critical difference	1.2439
MCC			
	Values		
N	5		
Chi-square	10.5333		
Degrees of freedom	4		
p value less than	0.0324		
5% critical difference	1.5626		

Table 4.5 Friedman test mean ranks (critical differences in boldface)—adaptive goal function of ACDT

Measure		acc	rec	prec	F1	MCC
Goal function	acc	**1.8333**	2.7500	2.9167	2.3333	2.5000
	rec	4.6667	**1.0000**	5.0000	3.6667	4.5000
	prec	3.6667	5.0000	**1.0000**	4.6667	3.6667
	F1	2.2500	2.8333	3.6667	**1.3333**	2.3333
	MCC	2.5833	3.4167	2.4167	3.0000	**2.0000**

Abbreviations: *acc* accuracy; *rec* recall; *prec* precision; *F1* F-measure; *MCC* Matthews correlation coefficient

In case of the remaining measures, the results do not always exceed the critical difference. Better, though not critically so, ranks are most often achieved for the combination of classification accuracy with F-measure and MCC.

4.4 Conclusions

The ACDT algorithm with five different measures for evaluating the classification quality, which are used as the goal function, enables improvement in the selected classification evaluation. For all of these approaches, it has been observed that change in the goal function leads to maximization of the selected measure, although the maximum value is sometimes found when a different measure is used. This possibility is of vital importance for data sets that are difficult to evaluate by using the classification accuracy measure. The proposal by the authors of [2] enables construction of trees based on division criteria that are related to accuracy, and yet to maximize them in terms of another, selected measure. This enables construction of a classifier which is not only accurate but also, for example, precise.

The results indicate that for ACDT the F-measure can be regarded as a better universal measure (i.e., one which yields better average result values) than accuracy. Further research is also required to evaluate the goal function in terms of the sizes of the constructed decision trees. The results of the experiments show that trees with a smaller number of nodes and of a smaller height are constructed when the recall measure is used, while larger trees are constructed when the precision measure is used.

References

1. J. Kozak, U. Boryczka, Dynamic version of the ACDT/ACDF algorithm for h-bond data set analysis, in *ICCCI*, pp. 701–710 (2013)
2. J. Kozak, U. Boryczka, Goal-oriented requirements for ACDT algorithms, in *International Conference on Computational Collective Intelligence* (Springer International Publishing, 2014), pp. 593–602
3. L. Rokach, O. Maimon, *Data Mining With Decision Trees: Theory and Applications* (World Scientific Publishing, 2008)

Chapter 5
Examples of Practical Application

5.1 Analysis of Hydrogen Bonds with the ACDT Algorithm

Hydrogen bonds (H-bonds) play a key role in both formation and stabilization of protein structures. They form and break while a protein deforms, for instance during the transition from a non-functional to a functional state. The intrinsic strength of an individual H-bond has been studied from an energetic point of view, but energy alone may not be a very good predictor [5].

The aim of the considered approach is to train ant agents to predict the stability of H-bonds in every protein. This is why we need to design an algorithm that analyzes complex data sets characterized by irregularly numerous decision classes. The effect of algorithm operation is improvement in classification accuracy for objects belonging to a specific decision class. Moreover, such improvement should have no impact on the classification precision.

This kind of combination of ant colony optimization algorithms with algorithms used for constructing decision trees was proposed in order to, for example, construct multiple different decision trees. Application of the pheromone trail gives additional possibilities of using the proposed algorithm when working with specially generated subsets of a decision table (tables generated in a fixed way from a single data set). In such a case, ant agents construct decision trees adapted to a given data subset by making use of trees generated earlier for a different data subset. This allows for better classification of objects that were originally difficult to extract. Such a solution is not a projection of any of the popular committees of classifiers, which are discussed in part II of this book. Though the method of sampling with replacement is used (bagging and boosting), it is implemented in a different way. Non-deterministic selection of the node split (random forests) was also introduced. The selection is not entirely random, because it is based on a heuristic and a reward for a given split, defined on the basis of the pheromone. Most importantly—the result of the algorithm is a single classifier (not an ensemble) constructed on the basis of multiple data sets.

© Springer International Publishing AG, part of Springer Nature 2019
J. Kozak, *Decision Tree and Ensemble Learning Based on Ant
Colony Optimization*, Studies in Computational Intelligence 781,
https://doi.org/10.1007/978-3-319-93752-6_5

Such a solution is used in case of this approach, because an earlier analysis (described, for example, in Sect. 3.4.1) indicated a big space of possible solutions. For all independent algorithm runs, the best decision trees obtained often present, despite very similar object classification accuracy, different confusion matrices. Thus when seeking for the decision tree which allows for the best classification, we find a great number of local optima in the solution space—which correspond to decision trees constructed in a different manner, but yielding the same classification quality.

5.1.1 Characteristics of the Dynamic Test Environment

ACDT has been adapted to a real-world data set concerning hydrogen bonds [5]. After discretization carried out with adjustment to ACDT in mind, the data set includes objects described by 35 condition attributes (with four values each and 1 decision attribute (with the 2 values denoted by "0" and "1"). The analyzed data was divided into three data sets: training, testing and clean data (described in detail in Table 5.1). Depending on the conducted experiments, the learning data differed with respect to the number of objects and the cardinality of decision classes, but the clean data represented always the same, unchanged data set.

We present a new approach involving creation of special pseudo-samples. The goal is to obtain better classification accuracy for objects belonging to decision class "0", with minimal cost function. This combination of high classification accuracy for objects from class "1" and small deterioration in precision for the second "0" class is used in the cost function.

In order to cope with such data, the data set was divided into pseudo-samples (see Fig. 5.1). The objects for the pseudo-samples were chosen so that the sample would contain a predetermined number of objects from each of the decision classes. Finally, the prepared data sets were used in the following stages of analysis in accordance with the rules presented below:

- Construction of decision trees: pseudo-samples.
- Testing of the trees: test set.
- Result: clean set.

Table 5.1 Data set used in experimental study

Training set	46 639 objects
	4.5% objects represent decision class 0
	95.5% objects represent decision class 1
Test set	126 151 objects
	5.0% objects represent decision class 0
	95.0% objects represent decision class 1
Clean set	111 143 objects
	6.1% objects represent decision class 0
	93.9% objects represent decision class 1

Fig. 5.1 Method of constructing pseudo-samples

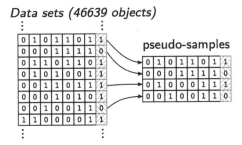

Data sets (46639 objects)

pseudo-samples

As the goal was to improve the quality of classification for class "0", the following data subsets (called pseudo-samples above) were prepared:

- **Learn 46,600**—all objects from the training set (46,639 objects) are used, of which 4.5% belong to decision class "0", and 95.5% to class "1".
- **Learn 3300**—one pseudo-sample consisting of randomly chosen roughly 3300 objects, of which 60% belong to decision class "0", and 40% to class "1".
- **Learn 4150**—one pseudo-sample consisting of randomly chosen roughly 4150 objects, of which 60% belong to decision class "0", and 40% to class "1".
- **Learn 8700**—one pseudo-sample consisting of randomly chosen roughly 8700 objects, of which 22% belong to decision class "0", and 78% to class "1".
- **Learn 3300, 4150 and 8700**—dynamic ACDT is run with randomly chosen pseudo-samples (in turn: **Learn 3300**, **Learn 4150** and **Learn 8700**).
- **Learn 3300, 4150 and 8700 V2**—dynamic ACDT is run with three pseudo-samples used sequentially with a greater impact of pruning.
- **Learn 15 sets**—dynamic ACDT algorithm in which 15 pseudo-samples with 3200 randomly chosen objects are used, where 50% objects belong to decision class "0", and 50% to class "1".
- **Learn 15 sets V2**—dynamic ACDT algorithm in which 15 pseudo-samples with 3200 randomly chosen pseudo-samples are used. There are different numbers of objects representing decision classes "0" and "1", respectively: 60–40%, 55–45%, 50–50%, 45–55%, 30–60%.

Where each time random choice is carried out by draw without replacement.

5.1.2 Conclusions

The results presented in Table 5.2 are the average values for 100 algorithm runs, each of them producing 15 generations for a population of 10 ant agents. It should be emphasized that during the algorithm run a two-part split criterion was applied. Additionally, ERB (Error-Based Pruning) was employed, being an extension of the pessimistic pruning classically used in ACDT. The introduced extension was mainly

Table 5.2 Results of the analysis of hydrogen bonds with applications of the ACDT algorithm

Data set	Accuracy for class 0	Accuracy for class 1	No. of nodes	Precision	No. of obj. from 0 class
Learn 46,600	0.0580	0.9937	180.4	0.3748	1046
Learn 3300, 4150 and 8700	0.3444	0.9533	310.1	0.3234	7205
Learn 3300, 4150 and 8700 V2	0.2806	0.9700	39.3	0.3774	5029
Learn 3300	0.6271	0.8545	257.1	0.2184	19429
Learn 4150	0.4847	0.9160	212.3	0.2721	12052
Learn 8700	0.3314	0.9576	254.1	0.3362	6669
Learn 3300 and 4150	0.4980	0.9018	271.3	0.2473	13624
Learn 15 sets	0.7309	0.7738	241.6	0.1732	28558
Learn 15 sets V2	0.7283	0.7748	267.2	0.1733	28431

based on the use of an additional parameter which allows for either increasing or decreasing the size of the classification error introduced during pruning.

ACDT shows a high capability of adapting to a dynamic environment (training set varying during algorithm operation). This can be seen from its fast adjustment to the current data set, while maintaining the information about the previous training set (in the form of pheromone trail). The results of experiments imply the possibility of improving classification accuracy for objects belonging to decision class "0" while maintaining similar precision and successful application of ACDT in a dynamically changing environment (solution space).

The receiver operating characteristic (ROC) (Fig. 5.2) allows us to see improvement in classification accuracy for objects from class "0", with a relatively small decrease in classification accuracy for objects from class "1", especially during the use of pseudo-samples Learn 330, 4150 and 8700 V2; Learn 3300, 4150 and 8700 and Learn 8700. Taking also into account the number of nodes in the decision trees, the algorithm yielded especially good results when working with Learn 3300, 4150 and 8700 V2 Learn 3300, 4150 and 8700 V2' i.e. when constructing decision trees for changing decision tables consecutively, with simultaneous reinforced decision tree pruning. The number of nodes in the decision tree is given in Table 5.2.

The precision of decision trees constructed by ACDT for decision class "0" equals 37% (Table 5.2). Though the classical selection of the entire training set for constructing the decision tree is reflected by a very high precision, we should note that in this case the total number of objects classified to decision class "0" is very low. The more attractive option is to obtain the precision of 37.74% during the decision tree construction using the learning data Learn 3300, 4150 and 8700 V2. In this partic-

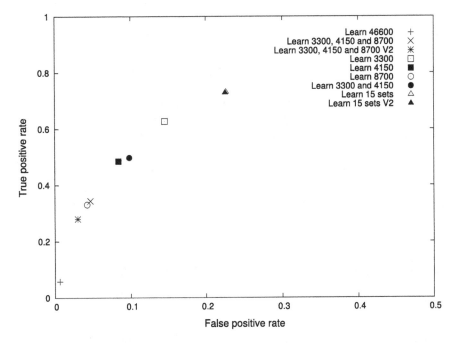

Fig. 5.2 The ROC curve for the results of dynamic ACDT

ular case, the number of properly classified objects from class "0" is much higher (classification accuracy for class "0" is about 28%) than in case of using the single, whole learning table (classification accuracy for class "0" of about 5%).

5.2 ACDT Algorithm for Solving the E-mail Foldering Problem

A special case of a classification problem is the e-mail foldering problem (EFP). It concerns a situation where e-mail users create new folders, and at the same time can stop using some of the folders created earlier. Additionally, the folders do not always correspond to the subjects of e-mails. Sometimes they can contain information about a task to be performed, group projects, or some of the recipients—and others make sense only with relation with previous messages.

The e-mail foldering problem is very complex because an automatic classification method that works well for one user can lead to mistakes for another one. Moreover, information can appear at different times, which causes additional difficulties. Analysis of contacts between the sender and the recipient, as well as the tree structure of folders created in every mailbox could contribute to solving this problem.

A similar problem was dealt with in [1], where the authors presented a case study of benchmark e-mail foldering based on the example of the Enron e-mail data set. R. Bekkerman et al. classified e-mails from seven e-mail boxes into thematic folders based on four classifiers: maximum entropy, naive bayes, support vector machine and wide-margin winnow.

Since that time, there have been multiple versions of ACDT used for solving the e-mail foldering problem. In most cases, solutions to that problem involved significant modifications to the basic version described in Chap. 3 and combination with other data mining methods. This chapter provides characteristics of the three approaches to the EFP problem described in [2–4].

5.2.1 Hybrid of ACDT and Social Networks

The authors of [2–4] each time based their approach on data preprocessing. The set of e-mail messages (the Enron e-mail data set) analyzed by the algorithm, described in detail e.g. in [3], was transformed into a classical decision table. However, each time the information about the contacts between message senders and recipients, as well as other persons included in the data set, was preserved.

In the basic version of the algorithm presented in [2], the authors proposed modification of ACDT that consisted in strengthening the pheromone trail for the classifiers (constructed with the classical ACDT approach) which assigned new e-mail messages to folders according to the decisions of various persons included in the recipient fields. This, of course, applied to messages addressed to multiple persons. In the remaining cases, the algorithm operated normally. The results obtained indicated improvement compared to those described in [1], and allowed for further development of the algorithm.

As defined, a social network is a multidimensional structure that consists of a set of social entities and the connections between them. Social entities are individuals who function within a given network, whereas connections reflect various social relations between the particular individuals. The authors of [3, 4] proposed extending the previous version of the algorithm by hybridization of the original algorithm with social network analysis.

The authors of [3] give the following schema of the algorithm's operation (Fig. 5.3). The first step (Fig. 5.3a) taken to use this method for improving the accuracy of assigning e-mail messages to folders is creation of a social network based on the contacts between the senders and the recipients of e-mails obtained from the Enron data set. All users of the mailboxes obtained from the Enron e-mail data set constitute vertices of the network, while its edges are the connections between those users for which the frequency of interaction, i.e. the number of e-mails sent between the given individuals, exceeds 10. Then, during the analysis of this network, the key users (hubs) and their immediate neighbors are selected (Fig. 5.3b), thus creating groups of users within the network.

Fig. 5.3 Diagram of the algorithm described in [3]

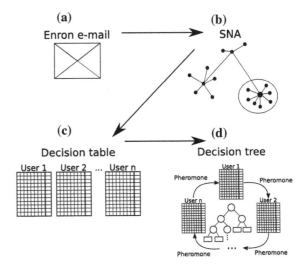

The next step taken in order to implement the proposed algorithm is transforming the set of data from the Enron e-mail data set into a decision table, separately for each mailbox within a given group (Fig. 5.3c).

While a classifier is being developed for each user, the communication network of the group assigned to that user is being analyzed. A classifier is built for a selected user of the algorithm, and then a new classifier is constructed for each subsequent individual in the group by using the same pheromone trail matrix.

After the classifiers have been built and the pheromone trails have stabilized, the final classifier for a given user is developed, in accordance with the diagram in Fig. 5.3d. This allows for retaining information on the decisions made by other members of the group (via the pheromone trail).

5.2.2 Conclusions

Authors of the algorithms pointed out that, according to Table 5.3, the results had improved significantly after creating a social network, based on which communication between the users had been analyzed. Every single time, the approach based on ACDT algorithm seemed to yield better results than the algorithms it was compared with. However, it was the last version of the algorithm, which included the social network, that in most cases allowed a better assignment of labels to the analyzed e-mail messages.

In [4], the authors presented further analysis and comparisons. Based on them, they concluded that development of a map of contacts in the form of a social network suggests that that the method will not only reduce the time spent on reading and replying to the e-mails received, but above all will be of crucial importance for

Table 5.3 Results regarding usefulness of the decisions assigning e-mails to folders

Data set	Classical algorithms presented in [1]				ACO algorithm presented in [2]	ACO algorithm presented in [3]
	MaxEnt	Naive Bayes	SVM	WMW		
beck-s	0.558	0.320	0.564	0.499	0.583	**0.600**
farmer-d	0.766	0.648	0.775	0.746	0.811	**0.834**
lokay-m	0.836	0.750	0.827	0.818	0.888	**0.891**
sanders-r	0.716	0.568	0.730	0.721	0.829	**0.871**
williams-w3	0.944	0.922	0.946	0.945	**0.962**	0.960
rogers-b	–	0.772	–	–	**0.911**	0.900
shackleton-s	–	0.667	–	–	0.709	**0.751**
steffes-j	–	0.755	–	–	0.841	**0.863**
symes-k	–	0.789	–	–	0.930	**0.937**

the process of information flow between company employees. It was noted that the proposed method of creating decision tables allows for the use of classical classifiers to categorize e-mail messages. However, the ACO based algorithm produces even better results due to its adaptability and the use of SNA elements.

5.3 Discovering Financial Data Rules with the ACDT Algorithm

Large opportunities for applying ACO based learning can be found in the financial data field. Research on the use of ACDT for predicting changes in the forex market is still underway. In this chapter we present a very interesting approach described in [7]. Other approaches to the use of decision trees for classification, like [6], are still being developed.

Thanks to the large volatility of currency pairs and the possibility of converting values, an international currency market has emerged over the last decades. Known as forex, it is the biggest market in the world, with a transaction volume reaching over 4.5 billion American dollars daily. From the viewpoint of a machine learning algorithm, the fast growth and the high daily volume have made the forex market an immense, often uncharted set of data.

The results given in [7] and presented in this chapter are even more interesting because they show ACDT in a new role. In this case, the problem that ACDT has to handle is not that of classical classification, but rather an association problem. In other words, feature selection is carried out, and the main problem is to find dependencies between the technical analysis indicators used in the forex market. The task of the described version of ACDT is to find a subset of those indicators that allow for further effective analysis of the market by a domain expert. The main difficulty is that the selected set of indicators should allow for effective prediction of the price movements while limiting the data set cardinality. This limitation is aimed at preventing data redundancy.

Due to the above, the ASC-ACDT algorithm is constructed in a manner allowing for predicting every feature (based on all remaining features, i.e. indicators). In case of good prediction quality, we can say that the feature under consideration, indicated as a decision attribute at the given time, is considered redundant.

Nowadays, a big potential for automatic transaction systems is becoming more and more visible. This applies to the forex transaction systems which include methods like High Frequency Trading (HFT), as well as the medium and long term trading approaches focused mostly on building effective portfolios. In both cases, a vast majority of the proposed methods are based on the concept that technical analysis indicators are capable of pointing out (with very high efficiency) the market instruments which should be the focus of interest for a potential decision maker. Rule definitions are given for example in [10]. In turn, in the review article [8] we can find information about failed approaches to technical analysis.

To the best knowledge of the authors of [7], there is no publication pinpointing the dependencies among the various technical analysis indicators. Thus ASC-ACDT may be considered as the first step towards reducing the complexity of decision rules for financial data. The described approach seems to be effectively scalable to the problem of multi-label analysis. At the same time, that approach may lead to discovering new transaction system rules.

5.3.1 ACDT for Financial Data

To adjust ACDT to financial data, the formal definition of the algorithm is changed. For the data set (3.2) and attributes A, the modified approach does not formally define any decision class. Each object in the set X can be described as:

$$x_i = ([v_i^1, \ldots, v_i^m]), v_i^j \in A_j, \tag{5.1}$$

where v_i^m is the value of attribute a_j for object x_i.

Then the ASC-ACDT algorithm can be formally defined as follows:

$$ASC\text{-}ACDT = \langle ((X, A), T(S), ants, p_{m, m_{L(i,j)}}(t), S \rangle. \tag{5.2}$$

The couple (X, A) represents the problem, with X being the set of objects, and A— the set of attributes, without any specially singled out decision attribute. Such a form of the algorithm allows for analyzing every single attribute in the same manner as the decision attribute in the original ACDT algorithm. Due to the specifics of the forex market financial data, the authors assume that the set of crucial indicators necessary for making a successful decision in that market can be reduced. Such a reduction of the set of attributes should lead to decreasing the problem complexity, while at the same time maintaining the prediction accuracy.

Due to specifics of the analyzed data, the authors are not only interested in the exact classification but assume that classification of an object to a neighboring decision class is also satisfactory. The assumption is that we treat as the decision attribute (denoted by i) the indicator whose value should be predicted. All decision classes (classes for attribute i) are sorted and labeled by consecutive numbers from 1 to N, where N is the overall number of such classes. The joint size of the neighboring classes is defined as:

$$neighbor_{size} = n\% \cdot N. \tag{5.3}$$

where the parameter n is supplied by an expert.

In order to cater for the inexact classification described above, the authors introduce a new fuzzy accuracy measure, which allows for measuring the efficiency of ASC-ACDT in case of a slight deviation from the target class:

$$d(DT, D)_{fuzzy} = \frac{TP^{-n\%} + TP + TP^{+n\%}}{|D|}, \tag{5.4}$$

Here $TP^{-n\%}$ is the number of objects classified to a neighboring class (preceding the target class in the adopted order of classes), while $TP^{+n\%}$ applies to a neighboring class following the target class in that order.

Figure 5.4 shows example classification, in which Fig. 5.4a contains the confusion matrix—with the assumption that all decision classes can be calculated and sorted. In this case, 10 classes with fuzziness equal to 0.1, i.e. 10%, can be seen. Thus classification to a neighboring class is considered a satisfactory result. All objects classified to the exact expected class are marked with a darker shade of gray, while those classified with a 10% error are marked with a lighter shade of gray.

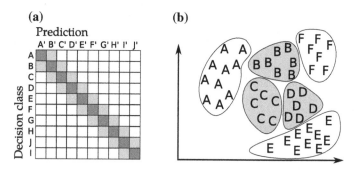

Fig. 5.4 Example of fuzzy accuracy used as a measure in the conducted experiments; **a** example confusion matrix, **b** example decision class C with satisfactory classification marked

The second part of Fig. 5.4b shows example solution for the selected class (in this example, C). Five different sets of objects classified to different decision classes can be seen. Additionally, the gray color marks the area containing classes which would be considered as satisfactory classification for an object from class C.

5.3.2 Conclusions

Kozak and Juszczuk [7] conducted experiments for the proposed modification. To deliver detailed results based on ASC-ACDT, the authors selected data which included 6 currency pairs: AUDCHF, CADCHF, EURCHF, GBPCHF, NZDCHF and USDCHF. For every currency pair, a time interval including 1500 readings was selected. The analysis involved discrete values of 10 technical indicators: Bears, Bulls, Average True Range (ATR), Demarket, Williams indicator, Commodity Channel Index (CCI), Relative Strength Index (RSI), standard deviation, momentum and Force Index (FI).

As stated by the authors, application of ASC-ACDT to prediction of indicator values in the forex market brings good results. The difficulty of prediction varies across different indicators: it is relatively easy to predict the values of Bears and Bulls, while the CCI indicator causes visible difficulties. However, the value of a given indicator can be almost always predicted with a similar probability of correctness. In the analyzed problem, all decision classes were considered to be enumerated types. Finally, as prediction of a value slightly different from the target one does not visibly affect the analysis, fuzzy accuracy was proposed as well—and the use of such accuracy turned out to allow for predicting indicator values with very good results. It should be noted that the above feature is as a rule observed in specific data only, and should not be treated as an universal conclusion.

Some of the results achieved by Kozak and Juszczuk are given in Table 5.4, to show that prediction of some indicators gives good results. The application of fuzzy accuracy can be clearly seen in Fig. 5.5, where the confusion matrix is shown for every analyzed case. The rows store information about the actual decision classes, while the columns store information on the predicted decision class (indicator value). A visible concentration of the values on the diagonal would mean error-free classification. It can be noticed that in most cases the prediction is very close to the diagonal—a darker color means a higher number of correct classifications.

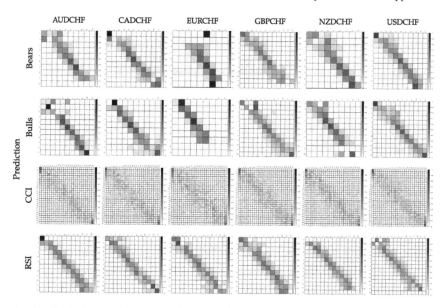

Fig. 5.5 Confusion matrix for the analyzed indicators. Darker grey color denotes higher number of objects assigned to the given decision class—rows store the actual value, and columns—the predicted indicator value

Table 5.4 Accuracy (acc.) and fuzzy accuracy (fuzzy acc.) of attribute prediction

Data set	CCI		RSI		Bears		Bulls	
	Acc.	Fuzzy acc.	Acc.	Fuzzy acc.	Acc.	Fuzzy acc.	Acc.	Fuzzy acc.
AUDCHF	0.2344	0.6915	0.4263	0.8520	0.5338	0.9333	0.5745	0.9512
CADCHF	0.2050	0.6747	0.3991	0.8406	0.5505	0.9369	0.5584	0.9497
NZDCHF	0.1969	0.6644	0.4129	0.8459	0.6354	0.9581	0.6656	0.9643
GBPCHF	0.2172	0.6959	0.4103	0.8394	0.4157	0.8200	0.4146	0.8311
EURCHF	0.2166	0.6921	0.3863	0.8309	0.5476	0.9535	0.5471	0.9593
USDCHF	0.2146	0.7078	0.3620	0.9427	0.4864	0.9031	0.5068	0.9035

5.4 Conclusions

The examples presented in this chapter confirm the adaptability of ACDT to specific problems, as well as its usefulness for supporting the process of solving those problems. It enriches the substantive analysis presented in Chap. 3 by the possibility of applying an ant colony optimization algorithm to constructing decision trees based on real-world data. Despite the fact that most of the applications consist in introducing some modifications to the canonical version of the algorithm, this should be treated as an advantage of the described algorithm.

The pheromone trail matrix, analyzed in detail in Sect. 3.4.1 and used in the approach from Sect. 5.1, allowed for learning one classifier based on multiple data sets used simultaneously. A similar solution (sometimes in a hybrid combination with social networks) was used in case of the algorithms described in Sect. 5.2. Application of ant colony optimization algorithms in the decision tree construction process was also proposed in [9]. The results of that paper are described in detail in Sect. 2.3. At the Cape Penisula University of Technology, there is ongoing research on applying ACDT (in the version presented in Chap. 3) to the intrusion detection problem. However, at this moment there is no information about the results obtained.

References

1. R. Bekkerman, A. McCallum, G. Huang, Automatic categorization of email into folders: benchmark experiments on enron and sri corpora. *Center for Intelligent Information Retrieval, Technical Report IR* (2004)
2. U. Boryczka, B. Probierz, J. Kozak, An ant colony optimization algorithm for an automatic categorization of emails, in *Computational Collective Intelligence. Technologies and Applications—6th International Conference, ICCCI 2014* (Seoul, Korea, 24–26 Sept 2014)
3. U. Boryczka, B. Probierz, J. Kozak, A new algorithm to categorize e-mail messages to folders with social networks analysis, in *Computational Collective Intelligence* (Springer International Publishing, 2015), pp. 89–98
4. U. Boryczka, B. Probierz, J. Kozak, Automatic categorization of email into folders by ant colony decision tree and social networks, in *Intelligent Decision Technologies 2016* (Springer International Publishing, 2016), pp. 71–81
5. I. Chikalov, P. Yao, M. Moshkov, J.C. Latombe, Learning probabilistic models of hydrogen bond stability from molecular dynamics simulation trajectories. BMC Bioinfo. **12**(S-1):S34 (2011)
6. P. Juszczuk, J. Kozak, K. Trynda, Decision trees on the foreign exchange market, in *Intelligent Decision Technologies 2016* (Springer International Publishing, 2016), pp. 127–138
7. J. Kozak, P. Juszczuk, Association ACDT as a tool for discovering the financial data rules, in *IEEE International Conference on Innovations in Intelligent Systems and Applications, INISTA 2017* (Gdynia, Poland, 3–5 July 2017), pp. 241–246
8. C. Park, S. Irwin, What do we know about the profitability of technical analysis? J. Econ. Surv. **21**(4), 786–826 (2007)
9. I. Surjandari, A. Dhini, A. Rachman, R. Novita, Estimation of dry docking duration using a numerical ant colony decision tree. Int. J. Appl. Manag. Sci. **7**(2), 164–175 (2015)
10. J.W. Wilder, New concepts in technical trading systems. Trend Res. (1978)

Part II
Adaptation of Ant Colony Optimization to Ensemble Methods

Chapter 6
Ensemble Methods

6.1 Decision Forest as an Ensemble of Classifiers

Many recent approaches to machine learning have benefited from the idea that predictions of an ensemble of models will usually be better than predictions of a single model. The errors of one model can be counteracted by the hits (in the context of incorrect predictions) of other models.

This collaborative approach to machine learning of ensembles can be implemented without interaction among the learning agents, which is known as ensemble learning, or with interaction during the learning stage, which is known as co-learning. To build an ensemble, we need to select a good method for constructing the individual classifiers included in its set.

Among the most frequent approaches to generating ensembles of classifiers, described for example in [9, 20, 32, 37], the methods recognized as fundamental are first of all bagging [6], boosting [34] and random forests [7].

All these approaches build ensembles by training each classifier on a bespoke data set. Boosting promotes diversity actively, whereas bagging relies on independent re-sampling from the training set. Boosting has been hailed the "best off-the-shelf classifier" by Leo Breiman himself, the creator of bagging. Both approaches construct an ensemble of any kind of classifiers, while random forests consist of decision trees only.

The main idea of the ensemble methodology is to combine a set of models, each of them solving the same original task, in order to obtain a better composite global model with more accurate and reliable results or decisions than can be obtained with a single model. The idea of building a predictive model by integrating multiple models has been under investigation for a long time. An ensemble of classifiers is a set of classifiers whose individual predictions are combined in some way (typically by voting) to classify new examples. One of the most active areas of research in supervised learning has been to study methods for constructing good ensembles of classifiers.

© Springer International Publishing AG, part of Springer Nature 2019
J. Kozak, *Decision Tree and Ensemble Learning Based on Ant Colony Optimization*, Studies in Computational Intelligence 781,
https://doi.org/10.1007/978-3-319-93752-6_6

Different classifiers are generated by manipulating the training set (as is done in boosting or bagging), manipulating the input features, manipulating the output targets or injecting randomness into the learning algorithm. The generated classifiers are then typically combined by either majority voting or weighted voting. Another approach is to generate classifiers by applying different learning algorithms (with heterogeneous model representations) to a single data set. Nowadays, data set sizes are continually increasing, as more and more information is stored electronically.

A decision forest, viewed as an ensemble of classifiers, is a collection of decision trees [7, 9, 32] given by the formula:

$$DF = \{d_j : X \rightarrow \{1, 2, \ldots, g\}\}_{j=1,2,\ldots,J}, \tag{6.1}$$

where J is the number of decision trees d_j $(J \geq 2)$.

In decision forests, predictions of the component decision trees are combined to make the overall prediction for the forest. Classification is carried out by a voting procedure. For example, in simple voting, each decision tree votes on classification of the sample, and the decision with the highest number of votes is chosen. The classifier determined by the decision forest (ensemble of classifiers) DF, denoted by $dDF : X \rightarrow \{1, 2, \ldots, g\}$, uses the following voting rule:

$$dDF(x) := \underset{c}{\operatorname{argmax}} N_c(x), \tag{6.2}$$

where c is a decision class, with $c \in \{1, 2, \ldots, g\}$, and $N_c(x)$ is the number of votes for classification of the sample $x \in X$ to class c, defined as $N_c(x) := \#\{j : d_j (x) = c\}$.

6.2 Bagging

Bagging is a set of classifiers based on bootstrap aggregation. It is commonly described as a one of the first, and at the same time the simplest set of classifiers. The approach was officially proposed by Breiman [6] (and described in 1994 [5]). Bagging is used to lower the variance and to decrease overtraining, which increases the stability of classification, at the same time improving its accuracy. The method is based on bootstrap aggregation [16] and on multiple construction of classifiers (mostly decision trees) on the basis of data subsets (bootstrap samples) generated from the whole learning set.

All data subsets (pseudo-samples) have exactly the same cardinality as the initial learning set (learning sample), and every sample is created through selection by draw. Assuming that the learning sample consisted of n elements, exactly n elements are selected for every pseudo-sample—whereby every case in the learning sample is selected with exactly the same probability equal to $\frac{1}{n}$. The described process is presented in Fig. 6.1 and as Algorithm 3.

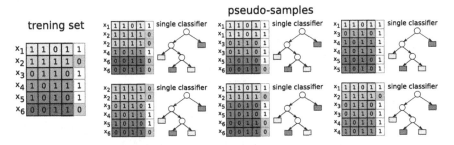

Fig. 6.1 Example of the bagging algorithm

Algorithm 3: The bagging algorithm

---

1 ensemble = *NULL*;
2 **for** number_of_classifiers **do**
3   // Construction of the classifier
4   data_set_classifier = choose_objects(data_set); // bootstrap aggregating
5   new_classifier = build_classifier(data_set_classifier);
6   ensemble.add(new_classifier);
7 **endFor**
8 result = ensemble;

Every prepared data set is used as a basis for constructing an independent, standard decision tree. This allows for deriving as many decision trees as there are pseudo-samples prepared in the way described above. Additionally, use of pruning is not approved, because in this particular case any possible overtraining of a single decision tree is naturally counteracted in the voting process by the remaining trees. The mentioned voting is the final classification process, described in detail at the beginning of this section. Each of the generated classifiers is granted exactly a single vote (no weights are included in the standard approach), and the set under consideration is assigned to the class which has obtained the greatest number of votes (Eq. 6.2).

6.3 Boosting

Boosting builds a set of classifiers on the basis of weighted observations. It can be considered as an alternative for bagging, and though the development of these two methods was independent, some similarities between them can be observed. Boosting was proposed in 1990 [34] by Schapire, who was inspired by Kearns [23]. Speaking more precisely, it was a proposal stemming from the following concept: a good learning set can be obtained from weak learning sets. A newer boosting method, namely discrete and adaptive boosting—known as AdaBoost, was introduced in 1995, and presented by Freund and Schapire e.g. in publications dating from 1996

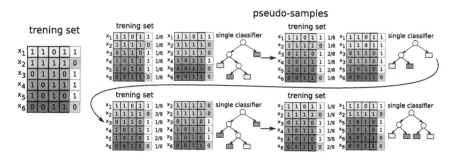

Fig. 6.2 Example of the boosting algorithm

and 1997 [18, 19]. The algorithm is still under development, which was shown for example in [33, 35].

In the boosting method, similarly as in case of bagging, a number of weak classifiers are built on the basis of specially generated pseudo-samples, and the classification process employs voting based on the derived classifiers. The pseudo-samples are created using draw with return. Namely, from an n element learning sample, a set of n elements is randomly selected for every pseudo-sample. The main difference between boosting and bagging is the probability distribution of selecting a given element from the learning set. Only in the first step the pseudo-sample is built exactly like in case of bagging, i.e. the probability of selecting every single element is equal exactly to $\frac{1}{n}$. Next, pseudo-samples are built depending on the way of classifying objects based on the decision tree (or other classifier) created from the first pseudo-sample. However, boosting does not require the use of re-sampling, and the construction of classifiers can be based on weighted observations.

Every element of the learning set is assigned a weight (initially equal for all objects) which determines the prescribed probability of selecting that element for the pseudo-sample being generated. After generating the pseudo-sample, a classifier is built on its basis and tested. The weight of all elements that have been classified on its basis to the wrong class is increased. This also increases the probability of selecting those objects for generating the next pseudo-sample. Other classifiers for the ensemble of classifiers under construction are built in a similar way—by modifying in every case the weights of elements of the learning set. Such an approach should in some way bring balance based on single classifiers so that the whole set of classifiers acquires better classification accuracy than that of a single decision tree constructed in the usual way (or other classifier). The described process is presented in Fig. 6.2.

In case of AdaBoost (see Fig. 6.4 for its high-level pseudocode), the weights of objects in the learning set are different. They are modified depending on the classification errors of the individual classifiers that are already included in the ensemble. Weight modification depends on the total weight of wrongly classified objects:

$$\varepsilon(j) = \sum_{x_i} we_i[c_i \neq c_i^j], \tag{6.3}$$

where we_i is the weight of object x_i, and c_i^j is the decision class of the analyzed object.

Such a modification is performed when the classification error is smaller or equal to 0.5. In the opposite case, the weights are multiplied by the coefficient (6.4), and then normalized.

$$\kappa(j) = \frac{1 - \varepsilon(j)}{\varepsilon(j)}. \tag{6.4}$$

In case of boosting, there is no simple voting (Eq. 6.2), but voting with weights, in which a single classifier gets the weight:

$$dDF(x) := \operatorname*{argmax}_{y \in Y} \sum_{j=1}^{J} \left(log \frac{1}{\kappa_j} \right) h_j(x, y). \tag{6.5}$$

Algorithm 4: The AdaBoost.M1 algorithm

---
1 weight_of_objects[1..n] = $\frac{1}{n}$;
2 ensemble = *NULL*;
3 weight_of_objects = *NULL*;
4 **for** number_of_classifiers **do**
5   // Construction of the classifier with a weighted vote
6   data_set_classifier = choose_objects(data_set, weight_of_objects);
7   new_classifier = build_classifier(data_set_classifier);
8   new_classifier.determine_the_weight_of_voting();
9   ensemble.add(new_classifier);
10   // Calculate the new weight of objects
11   **for** number_of_objects **do**
12   κ = classifies_object(current_object, ensemble); // by Eq. (6.4)
13   weight_of_objects[current_object] = weight_of_objects[current_object] $\cdot \kappa$;
14   **endFor**
15 **endFor**
16 result = ensemble;

Boosting is a classifier ensemble which has been extensively studied and has triggered numerous proposals for extending the original concept. For more information about that method, the reader is referred e.g. to Meir and Rätsch [28] and to multiple publications by Bühlmann and Hothorn [9, 10]. A review of the newest publications can be found on Schapire's website [36].

6.4 Random Forest

A good alternative to boosting can be found, for example, in [7], where Breiman proposed random forests. Despite the simplicity of constructing random forests, they still yield results very similar to the more complicated boosting [8]. Random forests can be seen as an improvement to bagging developed by Breiman based on comparisons with bagging and boosting. It should also be noted that, as opposed to the two ensembles of classifiers described before, in case of random forests some classifiers have to be used as decision trees.

The first stage of random forest construction is creation of pseudo-samples, based on which decision trees are later developed. The manner of selecting pseudo-samples is analogical to that used for bagging. Thus, out of an n-element set, n elements are drawn with return for every pseudo sample, whereby every object can be drawn with an identical probability equal $\frac{1}{n}$. Next, based on every pseudo-sample, a decision tree is built. This is done as many times as there are elements assigned to a single decision tree in every node. Consequently, the trees are generated without pruning.

The most important of the above stages is construction of the decision tree. In contrast to the standard method of building that tree, in case of random forests, during selection of the test for every node (separately), a random selection of attributes is carried out. Those attributes will be taken into consideration when generating the split at that node. Accordingly, every division is carried out on the basis of a different set of attributes in such a way that, independently of other randomization, m attributes (later taken into consideration during selection of the division) are drawn without return from all possible p features of objects in the learning set. An important problem is selection of the value for parameter m.

It is usually assumed that $m \ll p$, and experiments show that good results are obtained for $m = \sqrt{p}$. The described process is presented in Fig. 6.4 and in the form of Algorithm 5. In turn, a more detailed diagram of decision forest construction is shown in Fig. 6.3, to facilitate the interpretation of one of the versions of ACDF based on the random forest algorithm.

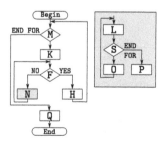

F- the completed tree
H- calculate the accuracy rate
K- create the pseudo-samples
L- create the subset of attributes
M- for each decision tree
N- select the next division
O- calculate the splitting rule value
P- select the best division
Q- the decision forest
S- for pseudo-samples and subset of att

Fig. 6.3 Diagram of random forest construction

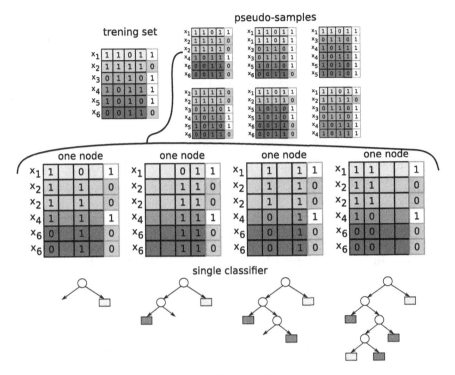

Fig. 6.4 Exemplary operation of the random forest algorithm

Decision trees constructed on the basis of a limited number of attributes have
visibly smaller sizes than trees built on the basis of all considered attributes. Random
forests can be used in case of more complex data sets. It should also be emphasized
that in case of random forests (as well as bagging, presented earlier) a non-biased
score of the probability of wrong classification is possible. Using the known fact that
$\frac{1}{3}$ objects are not selected to any given pseudo-sample (according to the probability
of their selecting equal $(1 - n)^n \approx e^{-n}$), and so $\frac{1}{3}$ trees in the forest are built without
any such object (similarly as in case of bagging), Breiman proposed the use of
observations not selected from the pseudo-sample of the given tree to check the
correctness of the classification by that tree. In such a way, after deriving the scores
for all trees we acquire an unbiased probability estimator of the wrong classification
by the random forest [7, 8].

Algorithm 5: Random forest algorithm

---

1 ensemble = *NULL*;
2 **for** number_of_classifiers **do**
3   // Construction of a decision tree classifier
4   data_set_classifier = choose_objects(data_set); // bootstrap aggregation
5   new_classifier = *NULL*;
6   **while** incomplete_decision_tree
7   attributes = create_subset_of_attributes($\sqrt{k}$ from all attributes);
8   division = select_next_division(data_set_classifier, attributes);
9   new_classifier.add(division);
10   **endWhile**
11   ensemble.add(new_classifier);
12 **endFor**
13 result = ensemble;

6.5 Evolutionary Computing Techniques in Ensemble Learning

A more detailed analysis of the applications of evolutionary computing techniques in data mining can be found in Chap. 2. This part is focused mainly on combining computational intelligence with ensemble learning. Once more, we are concerned first of all with the approach based on ant colony optimization.

One of the first approaches to ensemble learning based on ant colony optimization is the ant colony decision forest algorithm (ACDF). It is described in the further chapters of this book, and was published, depending on the version, e.g. in [3, 4, 25]. The basic version of ACDF dates from 2012, and was described in [3]. The authors proposed 7 different variants of ACDF, based mostly on bootstrap aggregation (known from bagging and the random forests). It should be emphasized that every version required application as a set of single classifier decision trees.

The details of that paper are described in Chap. 7.

In further papers Kozak et al. tried to improve ACDF by extending it based on the ideas known from boosting. In [4], an adaptive version of ACDF, named aACDF, was proposed. In aACDF, despite behavior similar to random forests, different weights for objects were applied following the selection of pseudo-samples (as described in Sect. 8.1). The current version of ACDF presented in Chap. 8 was published in 2015, in [25]. In that publication, Kozak et al. proposed, and compared with other approaches, two new versions of ACDF: self-adaptive ACDF (saACDF) and ACDF-Boost. They were connected even more with setting the weights for object selection to pseudo-samples, However, this time the above was done in an automatic way, without any additional parameters, as used previously in a ACDF.

Following ACDF, another application of ACO to ensemble learning, called eAnt-Miner, was proposed. This concept, described in [14], is based on building an ensemble (and more precisely, bagging), in which the individual classifiers are sets of rules obtained through applying Ant-Miner. As may be seen, in contrast to e.g. ACDF, eAnt-Miner is not a new type of ensemble method. The author justified the solution by the instability of results obtained with Ant-Miner. However, it should be pointed out that this is not a feature connected with operation of a stochastic algorithm. The instability problem was raised, for example, in [2], and some attempts at solving it were described in Sect. 2.2. Nevertheless, application of bagging together with the Ant-Miner actually improved the achieved results.

In the meantime, interesting ideas concerning applications of ACO to ensemble methods were proposed by Chen et al. and published, for example, in [11–13]. This was an interesting approach, in which the authors proposed building the classifiers simultaneously but with a different allocation of ant agents to building different classifiers in every single iteration. This means that in every iteration the agents were allocated to classifiers according to the equation proposed by the authors—which was the roulette wheel selection, known e.g. from genetic algorithms. In a 2014 paper [13], the authors presented three versions of ACO-Stacking, and compared them with the most popular ensemble methods, and additionally with approaches conceptually related to ACO-Stacking—stacking and GA-ensemble. In this case we should mention genetic algorithms based on stacking, as well as [26, 31].

Analogically to decision tree learning, also in case of attempts at ensemble learning based on evolutionary computing techniques, algorithms other than ACO were more often used. One of the first attempts to integrate evolutionary computing techniques with ensemble methods can be found, for example, in [22]. In that paper, already in 1999 the authors tried to create a hybrid of genetic programming and bagging or boosting, proposing application of the ensemble method to the creation of sub-populations. As it can be seen, this was in some sense an approach opposite to the idea chosen in this book (a more detailed review of the papers on GP, also as compared to ensemble methods, can be found in [17]). To combine this application with a genetic algorithm in the problem of optimizing the number of trees in a decision forest [1], the authors proposed a solution based on applying a genetic algorithm, in which the initial population consisted of properly selected decision trees. Finally, decision trees were used to test solutions with the use of bagging and random forests.

A combination with boosting (which is particularly useful e.g. in the analysis of the ACDF-Boost algorithm described in Sect. 8.3) can be found, for example, in the paper by Liu et al. [27], where a genetic algorithm is used in an ensemble approach based on single classifiers (decision rules). Another approach based on a genetic algorithm is proposed in [29], where the problem of optimizing the number of classifiers in an ensemble is discussed. Thanks to the genetic approach, the number of weak classifiers is reduced, and good classification accuracy is maintained as well. Another application of computational intelligence is discussed in a paper by Hidaka and Kurita [21]. The authors present a new application of particle swarm optimization (PSO) to ensemble classifiers. The concept proposed in that paper is to sequentially apply the classification algorithm to repeatedly modify the weights of objects.

6.6 Conclusions

This chapter introduces in a consistent way the subject of ensemble methods for classifiers , including in particular bagging, boosting (for the classifiers under consideration) and a random forest (only in case of decision trees). This is because bagging is in some sense the beginning of random forests, and these were the inspiration for the initial versions of ant colony optimization for learning an ensemble of classifiers, known as the ant colony decision forest. Boosting was the inspiration for the newest (published in 2015) version of ACDF [25].

We should note first of all the differences among those algorithms, and a kind of feedback that is observed in the boosting algorithm but does not appear in the other two methods—similarly as weighted voting. On the other hand, the random forest efficiently decreases the solution space, building at the same time many weak classifiers that globally cover most of the whole set of solutions.

The described methods have been extensively imitated, which helped popularize ensemble learning. In this case, it is certainly advisable to browse the popular publications on that subject, like, for example [15, 30, 41].

This was also seminal for development of other methods, as well as hybridization and emergence of such methods as: wagging or dagging, and other approaches, described e.g. in [24, 38, 39].

Additionally, besides the presented ensemble methods, considered as homogeneous ensembles, we should also note the concept of heterogeneous ensembles—especially the stacked generalization [40]. Moreover, a particularly recommended paper is the article paying special attention to the bias in random forest variable importance measures [37].

With regard to computational intelligence, all known ACO-based methods in ensemble learning have been presented. That part is supplemented with a short review of other evolutionary methods.

References

1. N. Adnan, Z. Islam, Optimizing the number of trees in a decision forest to discover a subforest with high ensemble accuracy using a genetic algorithm. Knowl.-Based Syst. **110**, 86–97 (2016)
2. U. Boryczka, J. Kozak, New algorithms for generation decision trees–ant-miner and its modifications, in *Foundations of Computational Intelligence*, vol. 6 (Springer, Berlin, Germany, 2009), pp. 229–264
3. U. Boryczka, J. Kozak, Ant colony decision forest meta-ensemble, in *International Conference on Computational Collective Intelligence* (Springer, Berlin, Heidelberg, 2012), pp. 473–482
4. U. Boryczka, J. Kozak, On-the-go adaptability in the new ant colony decision forest approach, in *Asian Conference on Intelligent Information and Database Systems* (Springer International Publishing, 2014), pp. 157–166
5. L. Breiman, Bagging predictors. Technical Report 421 (Depaxtment of Statistics, University of California at Berkeley, September 1994)
6. L. Breiman, Bagging predictors. Mach. Learn. **24**(2), 123–140 (1996)
7. L. Breiman, Random forests. Mach. Learn. **45**, 5–32 (2001)

8. L. Breiman, Random forests, www. available online
9. P. Bühlmann, T. Hothorn, Boosting algorithms: regularization, prediction and model fitting. Stat. Sci. **22**(4), 477–505 (2007)
10. P. Bühlmann, T. Hothorn, Twin boosting: improved feature selection and prediction. Stat. Comput. **20**(2), 119–138 (2010)
11. Y. Chen, M.L. Wong, An ant colony optimization approach for stacking ensemble, in *Second World Congress on Nature & Biologically Inspired Computing, NaBIC*, 15–17 December 2010, Kitakyushu, Japan (2010), pp. 146–151
12. Y. Chen, M.L. Wong, Optimizing stacking ensemble by an ant colony optimization approach, in *13th Annual Genetic and Evolutionary Computation Conference, GECCO 2011, Companion Material Proceedings*, Dublin, Ireland, 12–16 July, 2011 (2011), pp. 7–8
13. Y. Chen, M.L. Wong, H. Li, Applying ant colony optimization to configuring stacking ensembles for data mining. Expert Syst. Appl. **41**(6), 2688–2702 (2014)
14. G. Chennupati, Eant-miner: an ensemble ant-miner to improve the aco classification. arXiv:1409.2710 (2014)
15. T.G. Dietterich, Ensemble methods in machine learning, in *International Workshop on Multiple Classifier Systems* (Springer, 2000), pp. 1–15
16. B. Efron, Bootstrap methods: another look at the jackknife. Ann. Stat. **7**(1), 1–26 (1979)
17. P.G. Espejo, S. Ventura, F. Herrera, A survey on the application of genetic programming to classification. IEEE Trans. Syst. Man Cybern. Part C: Appl. Rev. **40**(2), 121–144 (2010)
18. Y. Freund, R.E. Schapire, Experiments with a new boosting algorithm, in *International Conference on Machine Learning* (1996), pp. 148–156
19. Y. Freund, R.E. Schapire, A decision-theoretic generalization of on-line learning and an application to boosting. J. Comput. Syst. Sci. **55**(1), 119–139 (1997)
20. J. Friedman, T. Hastie, R. Tibshirani, Additive logistic regression: a statistical view of boosting. Ann. Stat. **28**(2), 337–407 (2000)
21. A. Hidaka, T. Kurita, Fast training algorithm by particle swarm optimization for rectangular feature based boosted detector, in *Proceedings of FCV2008* (2008), pp. 88–93
22. H. Iba, Bagging, boosting, and bloating in genetic programming, in *Proceedings of the 1st Annual Conference on Genetic and Evolutionary Computation-Volume 2* (Morgan Kaufmann Publishers Inc., 1999), pp. 1053–1060
23. M. Kearns, Thoughts on hypothesis boosting, project for Ron Rivest's machine learning course at MIT (1988)
24. S.B. Kotsianti, D. Kanellopoulos, Combining bagging, boosting and dagging for classification problems, in *International Conference on Knowledge-Based and Intelligent Information and Engineering Systems* (Springer, 2007), pp. 493–500
25. J. Kozak, U. Boryczka, Multiple boosting in the ant colony decision forest meta-classifier. Knowl.-Based Syst. **75**, 141–151 (2015)
26. A. Ledezma, R. Aler, D. Borrajo, Heuristic search-based stacking of classifiers. Heuristics Optim. Knowl. Discov. **54** (2002)
27. B. Liu, B. McKay, H.A. Abbass, Improving genetic classifiers with a boosting algorithm, in *The 2003 Congress on Evolutionary Computation, CEC'03*, vol. 4 (IEEE, 2003), pp. 2596–2602
28. R. Meir, G. Rätsch, An introduction to boosting and leveraging, in *Advanced Lectures on Machine Learning, LNCS* (Springer, 2003), pp. 119–184
29. D. Y. Oh, *Ga-boost: A Genetic Algorithm for Robust Boosting* (University of Alabama, 2012)
30. D. Opitz, R. Maclin, Popular ensemble methods: an empirical study. J. Artif. Intell. Res. **11**, 169–198 (1999)
31. F.J. Ordóñez, A. Ledezma, A. Sanchis, Genetic approach for optimizing ensembles of classifiers, in *FLAIRS Conference* (2008), pp. 89–94
32. L. Rokach, O. Maimon, *Data Mining With Decision Trees: Theory And Applications* (World Scientific Publishing, 2008)
33. C. Rudin, R.E. Schapire, Margin-based ranking and an equivalence between AdaBoost and RankBoost. J. Mach. Learn. Res. **10**, 2193–2232 (2009)
34. R.E. Schapire, The strength of weak learnability. Mach. Learn. **5**, 197–227 (1990)

35. R.E. Schapire, The convergence rate of adaboost. COLT **10**, 308–309 (2010)
36. R.E. Schapire, Boosting, https://www.cs.princeton.edu/~schapire/boost.html
37. C. Strobl, A.L. Boulesteix, A. Zeileis, T. Hothorn, Bias in random forest variable importance measures: illustrations, sources and a solution. BMC Bioinf. **8**(1), 25 (2007)
38. G.I. Webb, Multiboosting: a technique for combining boosting and wagging. Mach. Learn. **40**(2), 159–196 (2000)
39. I.H. Witten, E. Frank, M.A. Hall, C.J. Pal, *Data Mining: Practical machine learning tools and techniques* (Morgan Kaufmann, 2016)
40. D.H. Wolpert, Stacked generalization. Neural Netw. **5**(2), 241–259 (1992)
41. Z.H. Zhou, *Ensemble Methods: Foundations and Algorithms* (CRC press, 2012)

Chapter 7
Ant Colony Decision Forest Approach

7.1 Definition of the Ant Colony Decision Forest Approach

Analysis of the operation of ACDT (with the most important observations presented in Chap. 3) gave rise to observations regarding the large diversity of solutions generated by independent runs of that algorithm. In particular, detailed research on the pheromone trail conducted with help of pheromone maps (see Sect. 3.3 for more information) inspired the idea of using ACDT to build multiple, independent decision trees, which can be joined together into an ensemble. The goal of such a solution is to generate more stable classifiers (classifier ensembles) which are capable of working with difficult to analyze and complex data sets.

The first results of research on this subject were published in [3], and are presented in this chapter. They concern mostly analysis of different solutions allowing for the use of ACO in ensemble learning (in particular, on the basis of ACDT). In general, those solutions are called the ant colony decision forest (ACDF), and are based on solutions known from the basic ensemble methods (like bagging and random forest [5–7]), but are adapted to using with ACO. Lately, in its newer versions, ACDF was extended to include adaptation, self-adaptation and use of the weights for examples in the data sets. Those solutions were published, for example, in [4, 9], and are described in Chap. 8.

Analogically to the ACDT approach described by Eq. (3.1), the ACDF approach can be represented as a six-tuple:

$$ACDF = \langle (X, A \cup \{c\}), Z, z, ants, p_{m,m_{L(i,j)}}(t), S \rangle, \qquad (7.1)$$

where:

$(X, A \cup \{c\})$—decision table representing the problem, where X is the set of objects and A the set of attributes, including the decision attribute c;

© Springer International Publishing AG, part of Springer Nature 2019
J. Kozak, *Decision Tree and Ensemble Learning Based on Ant
Colony Optimization*, Studies in Computational Intelligence 781,
https://doi.org/10.1007/978-3-319-93752-6_7

Z—set consisting of z decision trees built using ACDT with rules depending on the version of ACDF and based on equal votes (simple voting, Eq. (6.2))—solution to the problem,
including the following elements of an ant colony system:

$ants$—number of ants in an iteration,

$p_{m,m_{L(i,j)}}(t)$—solution selection rule for each node of the decision tree at time t,

S—set of acceptable objects in a node that indirectly performs the function of a taboo list.

It should be noted that the set of classifiers Z includes z trees built using rules imposed by ACDF. In many versions described below, those classifiers are interconnected and interdependent. However, there are cases, in which every classifier is built independently. The main feature of ACDF is the pheromone trail, which in the context of an ensemble operates as feedback not only for a single classifier, but also repeatedly for the whole set Z.

In this approach, for every classifier in the set Z, some subset of the decision table $(X, A \cup \{c\})$ is passed. That subset depends not only on the set of accepted objects S, but also on the way of building the pseudo-sample for every classifier. The method for building the pseudo-sample in turn depends on the model used in the given version of ACDF (the detailed manner of the model operation is described alongside every presented version of ACDF).

7.2 Ant Colony Decision Forest Based on Random Forests

A computational problem arises when an algorithm cannot guarantee finding the best hypothesis within the space of hypotheses. In the ACO and the random forest approaches, the task of finding a suitable hypothesis that fits the training data best is computationally intractable, so a more sophisticated method should be employed.

In case of ACDF based on random forests, ant agents create the collection of hypotheses in a random manner, complying with the threshold or the splitting rule. The challenge is to introduce a new random subspace method for growing collections of decision trees—which means that ant agents can create the collection of hypotheses from the hypothesis space using random-proportional rules. At each node of the tree, the ant agent makes a choice from a random subset (random pseudo sample) of attributes, and then constrains the tree-growing hypothesis to choose its splitting rule from that subset. Because of the re-labeled randomness proposed in this approach, the authors resign from choosing a different subset of attributes for each ant agent or colony in favor of greater stability of the adopted hypotheses. This is a consequence of the proposal first used in random forests.

ACDF is characterized by high diversity, because an ant agents make a cascade of choices which consist in choosing an attribute and its value at each internal node in the decision tree (to create a special hypothesis). Consequently, ensembles of

Table 7.1 Variants of the ACDF approach

Version	Tree selection mechanism	Data set selection	Variants of the set of attributes
ACDF_1 (Fig. 7.1a)	Local best trees	Individual, for ant agent	Individual, for ant agent
ACDF_2 (Fig. 7.1b)	Independently, global best trees	Individual, for ant agent	Individual, for ant agent
ACDF_3 (Fig. 7.1c)	Local best trees	Collective, for colony	Collective, for colony
ACDF_4 (Fig. 7.1d)	Independently, global best trees	Collective, for colony	Collective, for colony
ACDF-PH (Fig. 7.1e)	Independently, global best trees	Independently, for colony	Independently, for colony
ACDF-RF (Fig. 7.1f)	Local best trees	Collective, for colony	For each node, as in RF
ACDF-COOP (Fig. 7.1g)	Independently, global best trees	All samples	All attributes

decision trees perform better than individual decision trees due to the independently performed exploration/exploitation of the subspace of hypotheses.

This chapter presents various solutions regarding the manner of ACDF operation. Seven variants are selected, which differ in the way of preparing pseudo-samples and subsets of attributes, and choosing the best decision tree. The detailed operating manner of all considered variants is presented in Table 7.1 and Fig. 7.1.

All models of ACDF presented in this chapter are closely related to the random forest algorithm. The purpose of such an approach is to check how the idea of recording the solutions on the basis of the pheromone will work in combination with the continuous changes in the solution space (due to changing the analyzed cases and limitation on the number of attributes).

For a good analysis, see [3], which presents various versions of approaches to selecting decision trees for an ensemble of classifiers, as well as the manner of sharing pseudo-samples and attributes among ant agents.

In ACDF_1 each ant agent can search different subspaces of the solutions. In this case, a single tree is selected to the ensemble of classifiers from every iteration of the ant algorithm. This is the decision tree of the best quality (according (3.10)) from the given iteration. Such a choice ensures short algorithm operation time (in fact, the same as for ACDT). However, the classifiers in the ensemble are interdependent.

As the pheromone trail matrix, which provides a memory with the structure of the decision trees constructed earlier, acts towards copying the same solutions, classifiers in the ensemble can be very similar to each other. However, differentiation is ensured by solution space change, because an individual, independent pseudo-sample is generated for every ant agent (using bootstrap aggregation), and only a limited number of features are considered. Similarly to random forests, their number is the square

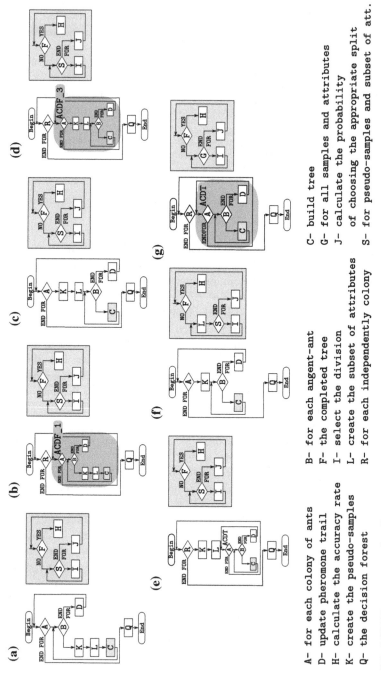

A- for each colony of ants   B- for each angent-ant   C- build tree
D- update pheromone trail   F- the completed tree   G- for all samples and attributes
H- calculate the accuracy rate   I- select the division   J- calculate the probability
K- create the pseudo-samples   L- create the subset of attributes   of choosing the appropriate split
Q- the decision forest   R- for each independently colony   S- for pseudo-samples and subset of att.

Fig. 7.1 ACDF forest construction diagram

root of the number of all features (attributes). However, in this case the subset of attributes is not selected in every node, but only a single time for every ant agent (and so a single classifier).

In ACDF_2, an identical way of generating the solution space was proposed (pseudo-samples and a subset of attributes). However, in this case the trees in the ensemble are mutually independent. Such a solution was obtained on the basis of multiple runs of ACDF_1, where each time the pheromone trail matrix was updated based on the initial value of the pheromone trail (Eq. (3.17)).

A smaller distribution in the hypothesis space for versions 3 and 4 is due to choosing the same sets of samples and attributes for each iteration of ants. As can be seen in Fig. 7.1c, d, elements 'K' and 'L' are in this case selected only once for any given iteration. This ensures larger cooperation among the ant agents, which within a single iteration operate on the same solution space. Similarly like in case of ACDF_1 and ACDF_2, ACDF_3 and ACDF_4 are different approaches built based on the same pheromone trail matrix. However, in case of ACDF_4 the decision trees included in the ensemble result from independent algorithm runs.

Another three versions of the algorithm have resulted from analysing the first approach to ACDF and allowing more distinction in the use of that solution (ACDF-PH and ACDF-RF), or introducing an innovation (ACDF-COOP).

ACDF-PH (ACDF-pheromone) is characterized by a smaller spatial distribution, so the pheromone values deposited in the subtrees play a more significant role. In this case, the authors wanted to obtain an algorithm based mostly on the pheromone trail, while preserving the solution space variability for each of the single classifiers added to the ensemble.

In turn, ACDF-RF (ACDF-random forests) is the closest reflection of the random forest algorithm, in which the subset of attributes is created at the moment of generating the split criterion at every single node. It should be pointed out that such an approach involves some sort of confrontation between the pheromone trail values (laid for different solution spaces) and the actual analyses related to every node with (perhaps) a different set of attributes. The trees added to the ensemble of classifiers are interdependent (similarly as in case of ACDF_1 and ACDF_3), while the pseudo-sample is set separately for every iteration of the algorithm .

The definitely most interesting approach (as compared to the classical ensemble methods) is the ACDF-COOP algorithm (ACDF-cooperation), which is not based on the idea of the random forest algorithm. In this particular case, the decision trees are created by ant agents independently, and only the inter-colony cooperation among the ant agents can be analyzed. Independent colonies have no possibility of communicating with each other, so a big diversity of decision trees can be obtained. An interesting fact is that every single selection of a node is performed using the whole learning sample (rather than the pseudo-sample) as well as all attributes.

Because of this, the single classifiers in the ensemble are not weak classifiers but single good decision trees (according to Eq. (3.10)), closely related to the multiple decision trees generated as a result of multiple runs of ACDF [10]. This is a novel approach compared to bagging, random forest or boosting.

The versions of ACDF with the greatest importance for this book are ACDF-PH (due to its close connection with the pheromone trail), ACDF-COOP (due to its high specificity—that approach is basically different from the known ensemble methods, e.g. because of the attempt to include the possibility of generating strong learners) and ACDF-RF (due to its high analogy to the random forest algorithm). For this reason, besides the diagram in Fig. 7.1, those approaches are presented additionally in the pseudocode form—see Algorithms 6, 7 and 8. This allows for their better comparison with other approaches described further on in this book. The above versions of ACDF are also subject to comparative analysis with other ensemble methods, described in Chap. 8.

Algorithm 6: The ACDF-PH algorithm

---

1 *meta_ensemble := null*;
2 **for** $j = 1$ **to** *number_of_classifiers* **do**
3   pheromone_trail_initialization(*pheromone*);
4   *best_classifier* := null;
5   *data_set_classifier* :=choose_objects(*data_set*); //equal probability
6   *classifier_attributes* :=choose_attributes(*set_of_attributes*); //equal probability
7   **for** $i = 1$ **to** *number_of_iterations* **do**
8   *local_best_classifier* := null;
9   **for** $a = 1$ **to** *ants_per_colony* **do**
10   // Classifier construction
11   *new_classifier* := build_tree_using_ACDT(*data_set_classifier*, *classifier_attributes, pheromone*);
12   **if** *new_classifier* **is_of_higher_quality_than** *local_best_classifier* **then**
13   *local_best_classifier* := *new_classifier*;
14   **endif**
15   **endfor**
16   pheromone_trail_update(*local_best_classifier, pheromone*);
17   **if** *local_best_classifier* **is_of_higher_quality_than** *_classifier* **then**
18   *best_classifier* := *local_best_classifier*;
19   **endif**
20   **endfor**
21   *meta_ensemble*.add(*best_classifier*);
22 **endfor**
23 *result := meta-ensemble*;

Algorithm 7: The ACDF-RF algorithm

---

1 pheromone_trail_initialization(*pheromone*);
2 *meta_ensemble := null*;
3 **for** *j* = 1 **to** *number_of_classifiers* **do**
4   *best_classifier* := null;
5   *data_set_for_classifier* :=choose_objects(*data_set*); //equal probability
6   **for** *a* = 1 **to** *ants_per_colony* **do**
7   // Classifier construction
8   // A subset of attributes is created at each node (ACDT_att)
9   *new_classifier* := build_tree_using_ACDT_att(*data_set_for_classifier, pheromone*);
10   **if** *new_classifier* **is_of_higher_quality_than** *best_classifier* **then**
11   *best_classifier* := *new_classifier*;
12   **endif**
13   **endfor**
14   update_pheromone_trail(*best_classifier, pheromone*);
15   *meta_ensemble*.add(*best_classifier*);
16 **endfor**
17 *result := meta-ensemble*;

Algorithm 8: The ACDF-COOP algorithm

---

1 *meta_ensemble := null*;
2 **for** *j* = 1 **to** *number_of_classifiers* **do**
3   pheromone_trail_initialization(*pheromone*);
4   *best_classifier* := null;
5   **for** *i* = 1 **to** *number_of_iterations* **do**
6   *local_best_classifier* := null;
7   **for** *a* = 1 **to** *ants_per_colony* **do**
8   // Classifier construction
9   // Classical ACDT algorithm - always all objects and attributes
10   *new_classifier* := build_tree_using_ACDT(*data_set, pheromone*);
11   **if** *new_classifier* **is_of_higher_quality_than** *local_best_classifier* **then**
12   *local_best_classifier* := *new_classifier*;
13   **endif**
14   **endfor**
15   pheromone_trail_update(*local_best_classifier, pheromone*);
16   **if** *local_best_classifier* **is_of_higher_quality_than** *_classifier* **then**
17   *best_classifier* := *local_best_classifier*;
18   **endif**
19   **endfor**
20   *meta_ensemble*.add(*best_classifier*);
21 **endfor**
22 *result := meta-ensemble*;

7.3 Computational Experiments

A variety of experiments were conducted to test the performance and behavior of ACDF. The most important results are presented in Table 7.3 and in the section on statistical tests. All experiments were repeated 30 times for each data set. Each experiment included 1250 generations, with the ant colony population size of 50. In each case, the decision forest consisted of 25 trees. A comparative study of ACDT (described in [1, 2]) was performed for the seven different versions of the algorithm described in this chapter to examine ant colony optimization ensemble learning based on the random forest. Moreover, comparisons with the ACDT-forest algorithm were carried out. The latter algorithm produces the set of decision trees resulting from a single run of ACDT (by choosing the best decision tree from every iteration). Such a solution was selected to compare the variants of ACDT with its most basic form. In turn, the comparison with classical methods was described in Chap. 8.

Evaluation of the performance of ACDF methods was carried out using 12 public domain data sets from the UCI (University of California at Irvine) data set repository. The main characteristics of those data sets are shown in Table 7.2. Like in case of the ACDT approach, data sets larger than 1000 objects were divided into the training and testing sets in a random way. Data sets with fewer than 1000 objects were estimated by 10-fold cross-validation. In both cases, an additional data set—a clean set—was also used. The results were tested on a clean set that had not been used to build the classifier.

It should be noted that the sets analyzed in this chapter can be distinguished from those selected for testing ACDT (Sect. 3.4) because the present ones have been selected in such away that most of the analyzed differences concern the use of ensemble methods. Moreover, results of a single run of ACDT can be in some

Table 7.2 Original parameters in data sets

Data set	Number of instances	Number of attributes	Decision class
heart	270	13	2
breast-cancer	280	9	2
balance-scale	625	4	3
dermatology	366	34	6
hepatitis	155	19	2
breast-tissue	106	9	6
cleveland	303	13	5
bcw	699	9	2
lymphography	148	18	4
shuttle	72,500	9	7
mushroom	8124	22	7
optdigits	6520	64	10

Table 7.3 Comparative study—accuracy rate

Data set	ACDT–tree		ACDT–forest		ACDF_1		ACDF_2		ACDF_3		ACDF_4		ACDF-PH		ACDF-RF		ACDF-COOP	
	Acc.	#n	Acc.	#n	Acc.	#n	Acc.	#n	Acc.	#n	Acc.	#n	Acc.	#n	Acc.	#n	Acc.	#n
heart	0.7744	13.0	0.7628	253.4	0.8269	121.8	0.8311	133.5	0.8147	120.2	0.8194	129.4	0.8191	155.1	**0.8391**	194.8	0.7781	257.8
breast-cancer	0.7165	6.5	0.7284	82.1	0.7363	68.0	0.7308	83.6	0.7367	74.2	0.7303	85.9	0.7313	92.7	**0.7414**	96.0	0.7308	122.9
balance-scale	0.7821	51.6	0.8003	1198.8	0.7877	440.5	0.8343	440.7	0.8209	416.5	0.8180	428.4	0.8299	426.9	**0.8559**	884.3	0.8202	1135.7
dermatology	0.9339	7.5	0.9314	113.7	0.9290	232.6	0.9122	225.0	0.9114	231.2	0.9033	225.0	0.9135	242.8	0.9690	297.7	**0.9765**	272.1
hepatitis	0.7989	5.0	0.8005	43.4	0.8190	48.6	0.8085	68.4	0.8089	60.1	0.8045	70.7	0.8152	77.8	**0.8210**	58.4	0.8145	79.9
breast-tissue	0.4702	12.3	0.4611	205.1	0.4810	138.3	0.4782	124.1	0.4769	135.4	0.4710	121.7	0.5029	151.7	0.4919	146.0	**0.5167**	198.5
cleveland	0.5401	15.7	0.5456	338.7	0.5660	133.1	0.5578	136.5	0.5549	128.4	0.5575	133.7	0.5590	165.7	0.5737	202.9	**0.5795**	291.4
bcw	0.9301	9.3	0.9306	103.2	0.9363	245.4	0.9441	131.8	**0.9544**	428.4	0.9514	207.4	0.9519	146.1	0.9280	275.1	0.9505	127.1
lymphography	0.7828	8.3	0.7821	161.6	0.8524	183.6	**0.8857**	116.0	0.8068	200.2	0.7857	102.2	0.7782	115.5	0.8265	318.5	0.8122	167.5
shuttle	0.9971	62	0.9975	1508	0.9969	3464	0.9965	5557	0.9960	7890	0.9957	7875	0.9934	7992	0.9972	2458	**0.9976**	1510
mushroom	0.6309	71	0.6313	1604	**0.6506**	654	0.6426	633	0.6383	604	0.6398	592	0.6420	579	0.6475	1007	0.6331	1617
optdigits	0.8021	151	0.8751	4273	0.9111	4222	0.8999	4112	0.8843	4236	0.8882	4157	0.8862	4725	**0.9442**	3942	0.9429	4235

Abbreviations: *Acc.* accuracy rate; *#n* number of nodes

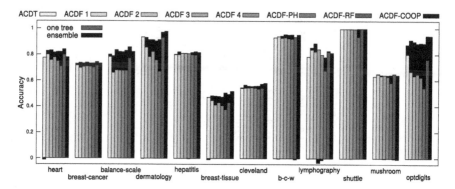

Fig. 7.2 Decision tree and decision forest accuracy rates

cases different from those presented in Sect. 3.4 because in the latter case an earlier version of ACDT was used, as well as different parameters, employed by the authors in [1–3].

7.3.1 Comparison of Different Versions of ACDF

In the most of the experimental studies, the seven variants of ACDF give a better performance than ACDT on a vast majority of data sets (see Table 7.3 and Fig. 7.2).

Most probably, the reason for the good results of ACDF is that the final classification is based on many independent searches of the solution space during the process of ant agent learning via pheromone values. This can be seen especially in case of ACDF-PH, where the individual decision trees are weak classifiers but the ensemble of such weak classifiers is very good at the accuracy of classification.

Figure 7.2 shows the classification accuracy of a single decision tree in an ensemble of classifiers (the best of all—grey column) compared to the classification accuracy of the whole decision forest (black column).

Analyzing the performance of ACDF-RF, we can observe that the results obtained are interesting—high classification accuracy compared to other analyzed versions of ACDF can be noted. The first observation in case of ACDF-COOP is that, judging by the results, it is better to use independent colonies (iterations) and separately analyzed pheromone tables. Such an approach usually gives better results in all cases (with regard to classification accuracy). This can be seen in Fig. 7.3, where the numbers of nodes are presented. As for the two methods: ACDT (with ACDT-forest) and ACDF-COOP, those techniques are characterized by the same number of nodes in the created decision trees. When comparing ACDT to the remaining versions of ACDF, we can see that the number of nodes is reduced. Analysis of the results in terms of classification accuracy—prediction of new samples—shows that the ACDF approaches are promising.

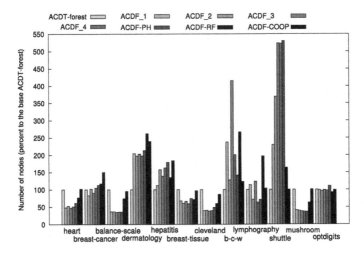

Fig. 7.3 Number of nodes (in percents compared to the base ACDT-forest)

The results in Table 7.3 suggest that performance of the presented ACDF approach is interesting and unaffected by local minima in the hypothesis space. Nevertheless, the results also show that the use of an ensemble of classifiers can mitigate the problem of diversity loss in the tree structure, which occurs in the obtained decision forest.

7.3.2 Statistical Analysis

The results of statistical analysis are presented in Tables 7.4 and 7.5. The non-parametric Friedman test confirms that ACDF is significantly better than ACDT and ACDT-forest. In Table 7.5, bold fonts indicate the values that meet the criterion of 5% critical difference (the smaller the value, the better the algorithm, statistically speaking). Table 7.4 presents all values determined during statistical analyses, including the mean ranks for samples.

We should note the ACDF-RF algorithm is much better than other approaches. With the range equal to 2.4167, it is critically better than the most of the analyzed approaches (the critical values were not exceeded only in case of comparison with ACDF_1 and ACDF-COOP). Again, a very interesting aspect is analysis of a novel approach based on generating an ensemble of classifiers consisting of very good single classifiers: ACDF-COOP is second in terms of the required range, often reaching the critical difference. It should be noted that ACDF_1 has reached a surprisingly high position, while ACDF_3 is only better (but not critically) than ACDT and the ensemble of decision trees resulting from the operation of ACDT (ACDT-forest).

Table 7.4 Friedman test
results and mean ranks (the
best rank in boldface)

Accuracy	
	Values
N	30
Chi-Square	38.454482
Degrees of freedom	8
p value is less than	0.0001
5% critical difference	1.795835
Mean ranks	
ACDT	7.6667
ACDT-forest	7.1667
ACDF_1	3.7500
ACDF_2	4.2917
ACDF_3	5.3333
ACDF_4	6.0833
ACDF-PH	4.7500
ACDF-RF	**2.4167**
ACDF-COOP	3.5417

7.4 Example of Practical Application

The ACDT algorithm was used for the hydrogen bonds analysis described in detail
in Sect. 5.1. In [8], the ACDF algorithm in the ACDF-RF version , ACDF_1 (but
without using subsets of attributes—features) and ACDF_3 (also without subsets of
attributes) were used for H-bond analysis, and eventually compared to the classical
ACDT.

Additionally, in conjunction with a specific, real-world problem involving two
decision classes, all three models are modified in such a manner that the decision
class of a smaller cardinality is strengthened. In this case, behavior of ACDF is similar
to that of boosting [11] with regard to performance. Each created pseudo-sample is
based on a weighted draw so that objects from a smaller number of participants have
a greater probability of being chosen to the pseudo-sample. In this way, decision
trees are created based on such pseudo-samples containing objects with the same
number of participants for class "0" and "1".

In comparison to ACDT (Sect. 5.1, Fig. 5.2), better results are obtained with the
use of ACDF, which is illustrated in the Fig. 7.4. In this case, a break through the
pareto front achieved by ACDF performance can be observed, of particular interest
is the ACDF-RF approach, in which an individual decision tree (treated as a result
of the forest) yields a very poor outcome. However, the set of such weak and small
decision trees produces better results, which was confirmed in [3]. Hence a good
classifier embodied by an ensemble of classifiers could be built out of those decision
trees (Table 7.6).

Table 7.5 Differences between ACDF algorithms (critical differences in boldface)

	Accuracy								
	ACDT	ACDT forest	ACDF 1	ACDF 2	ACDF 3	ACDF 4	ACDF PH	ACDF RF	ACDF COOP
ACDT	–	0.5000	**3.9167**	**3.3750**	**2.3333**	1.5833	**2.9167**	**5.2500**	**4.1250**
ACDT-for.	–0.5000	–	**3.4167**	**2.8750**	**1.8333**	1.0833	**2.4167**	**4.7500**	**3.6250**
ACDF_1	**–3.9167**	**–3.4167**	–	–0.5417	–1.5833	**–2.3333**	–1.0000	1.3333	0.2083
ACDF_2	**–3.3750**	**–2.8750**	0.5417	–	–1.0417	–1.7917	–0.4583	**1.8750**	0.7500
ACDF_3	**–2.3333**	**–1.8333**	1.5833	1.0417	–	–0.7500	0.5833	**2.9167**	**1.7917**
ACDF_4	–1.5833	–1.0833	**2.3333**	**1.7917**	0.7500	–	**1.3333**	**3.6667**	**2.5417**
ACDF-PH	**–2.9167**	**–2.4167**	1.0000	0.4583	–0.5833	–1.3333	–	**2.3333**	**1.2083**
ACDF-RF	**–5.2500**	**–4.7500**	–1.3333	**–1.8750**	**–2.9167**	**–3.6667**	**–2.3333**	–	–1.1250
ACDF-COOP	**–4.1250**	**–3.6250**	–0.2083	–0.7500	**–1.7917**	**–2.5417**	–1.2083	1.1250	–

Table 7.6 Use of ACDF for H-bond analysis

Method	Accuracy for class 0		Accuracy for class 1		N. of nodes		Precision		N. of obj. from 0 class	
	ens.	one tree	ens.	one tree	ens.	one tree	ens.	one tree	ens.	one tree
ACDF_1	0.6296	0.5747	0.9042	0.9009	9068	302	0.2905	0.2612	4211	3754
ACDF_3	0.6671	0.5945	0.8984	0.8954	8488	281	0.2985	0.2734	4315	3763
ACDF-RF	0.6783	0.0470	0.8745	0.9722	1987	12	0.2750	0.0475	4494	286

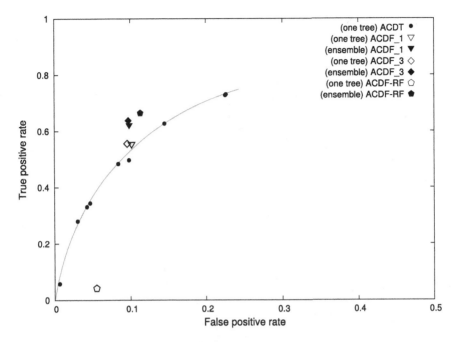

Fig. 7.4 The ROC curve for the results of ACDF

7.5 Conclusions

An ensemble of classifiers built with ACDF in the most cases yields better classification accuracy, while at the same time the number of nodes is reduced compared to the decision forests constructed by the classical ACDT algorithm (the set of best classifiers).

For versions of the algorithm based on decision forests, in case of ACDF the individual trees exhibited worse classification accuracy compared to the trees in ACDT. This, however, had a positive impact on the constructed decision forest, characterized by improvement in the classification accuracy. In turn, in the method inspired by the random forest algorithm, an interesting dependency was observed, which indicated that maintaining good classification accuracy of a single decision tree can improve the classification accuracy of the whole decision forest.

Statistical analysis also confirms that ensemble learning based on ACO and close to the random forests allows for obtaining better results than with other approaches. Detailed statistics point out that the larger the variety of decision trees in the ensemble, the better the quality of the whole ensemble of classifiers.

Also practical use of the selected ACDF models based on random forests allows us to conclude that an ensemble of classifiers based on ACO is better at analysing difficult data sets than a single classifier based on ACO. Analysis of selected variants related to the ACDF algorithm based, for example, on the idea of boosting and other classical ensemble methods is given in Chap. 8.

References

1. U. Boryczka, J. Kozak, Ant colony decision trees—a new method for constructing decision trees based on ant colony optimization, in *Computational Collective Intelligence*, Technologies and Applications, Volume 6421 of Lecture Notes in Computer Science, ed. by J.-S. Pan, S.-M. Chen, N. Nguyen (Springer, Berlin, Heidelberg, 2010), pp. 373–382
2. U. Boryczka, J. Kozak, An adaptive discretization in the ACDT algorithm for continuous attributes, in *Computational Collective Intelligence*. Technologies and Applications, Volume 6923 of LNCS (Springer, 2011), pp. 475–484
3. U. Boryczka, J. Kozak, Ant colony decision forest meta-ensemble, in *International Conference on Computational Collective Intelligence* (Springer, Berlin, Heidelberg, 2012), pp. 473–482
4. U. Boryczka, J. Kozak, On-the-go adaptability in the new ant colony decision forest approach, in *Asian Conference on Intelligent Information and Database Systems* (Springer International Publishing, 2014), pp. 157–166
5. L. Breiman, Bagging predictors. Technical Report 421 (Department of Statistics, University of California at Berkeley, September 1994)
6. L. Breiman, Bagging predictors. Mach. Learn. **24**(2), 123–140 (1996)
7. L. Breiman, Random forests. Mach. Learn. **45**, 5–32, October 2001
8. J. Kozak, U. Boryczka, Dynamic version of the ACDT/ACDF algorithm for h-bond data set analysis, in *ICCCI* (2013), pp. 701–710
9. J. Kozak, U. Boryczka, Multiple boosting in the ant colony decision forest meta-classifier. Knowl. Based Syst. **75**, 141–151 (2015)
10. J. Kozak, U. Boryczka, Collective data mining in the ant colony decision tree approach. Inf. Sci. **372**, 126–147 (2016)
11. R.E. Schapire, The strength of weak learnability. Mach. Learn. **5**, 197–227 (1990)

Chapter 8
Adaptive Ant Colony Decision Forest Approach

8.1 Adaptive Ant Colony Decision Forest

This chapter presents the adaptive ant colony decision forest algorithm (aACDF), which was first proposed in [2]. The aim of this approach is learning an ensemble of classifiers based on data sets created during the algorithm run. The aACDF approach allows for constructing classifiers in which wrongly classified objects obtain a greater probability of being chosen for pseudo-samples in the subsequent iterations. Every pseudo-sample is created on the basis of training data, and the samples are created for each population of ant agents. It is worth mentioning that creation of samples is based on the results obtained during the decision tree construction until that time. The motivation is improvement in the quality of classification by decision trees, which depends on the classification accuracy and the decision tree growth observed in the generated ensemble of classifiers.

The aACDF concentrates on samples which are weak or difficult to discriminate or classify. This is the idea of the boosting algorithm, which reduces sensitivity to training samples and tries to force the analyzed approach to change performance by constructing a new distribution over samples based on the results generated previously. That approach was first proposed by Schapire in 1990 [11] (and next described in [8], which was inspired by Kearns [9]). A detailed description of boosting can be found in Sect. 6.2—a noteworthy aspect there is the way of calculating the objects for pseudo-samples and setting their selection probability. This is especially useful from the viewpoint of the ACDF algorithms presented in this chapter. The ACDF-RF algorithm was a prototype for the adaptive ACDF algorithm described in Chap. 7, where the similarities to random forests are emphasized. In [2] the authors described a different method of generating pseudo-samples for each population of ant agents. The on-the-go, dynamic pseudo-samples are chosen according to the previously

© Springer International Publishing AG, part of Springer Nature 2019
J. Kozak, *Decision Tree and Ensemble Learning Based on Ant
Colony Optimization*, Studies in Computational Intelligence 781,
https://doi.org/10.1007/978-3-319-93752-6_8

obtained classification quality. The adaptability is focused on weak samples. Selection of objects is carried out by sampling with replacement from an n-element set, and the sample always consists of n objects.

The initial probability of object selection is equal to $\frac{1}{n}$ (like in bagging or random forests). In the next population of ant agents, the value of this probability depends on the weight of the given object. In case of incorrect classification, the weight coefficient is increased according to the formula:

$$we_i = \begin{cases} 1, & \text{if the object is classified well} \\ 1 + \lambda \cdot n, & \text{otherwise,} \end{cases} \qquad (8.1)$$

Meanwhile, the probability of selecting the object is calculated according to the formula:

$$pp(x_i) = \frac{we(x_i)}{\sum_{j=1}^{n} we(x_j)}. \qquad (8.2)$$

First of all, attention should be paid to the reinforcement applied to the objects (observations) that are badly classified by the current (at the given time) ensemble of classifiers. In other words, during generation of the first pseudo-sample, all objects have the same probability of being chosen (Eq. 8.2). Then, the objects that were not correctly classified by the generated classifier get some sort of reinforcement according to Eq. (8.1). After building another decision tree and adding it to the ensemble, there is another update of the weights for all considered objects. This whole process is repeated until an appropriate number of classifiers in the ensemble is obtained. The aACDF algorithm is presented as Algorithm 9.

An essential issue in case of aACDF is the value of parameter λ, determining the strength of reinforcement. A higher value means a bigger increase in the probability of selecting an object that was badly classified to the new pseudo-sample. However, if $\lambda = 0$, there is no feedback and the algorithm operates in the same way as ACDF-RF, while for the $\lambda = 1$ properly classified objects have the weight equal 1, and incorrectly classified objects the weight equal $1 + n$, where n is the cardinality of the learning set. In this case, the value of λ should be set in advance—depending on the analyzed data set. Experiments concerning the value of λ in aACDF are described in Sect. 8.4.1. In turn, a comparison of the aACDF algorithm with other approaches is presented in Sect. 8.4.2.

8.2 Self-adaptive Ant Colony Decision Forest

During the analysis of aACDF described in Sect. 8.1, we have noted that the greater support for incorrectly classified objects results in placing a larger number of these objects in the pseudo-samples, which allows for improving the quality of classification. However, use of the λ parameter requires earlier tuning of the algorithm. That is why in [10] Kozak et al. proposed a modification of aACDF known as the

---

Algorithm 9: Adaptive ACDF algorithm

---

1 pheromone_trail_initialization(*pheromone*);

2 *weight_of_objects[1..n]* := $\frac{1}{n}$;

3 *meta_ensemble* := *null*;

4 **for** *j* = 1 **to** *number_of_classifiers* **do**

5   *best_classifier* := null;

6   **for** *a* = 1 **to** *ants_per_colony* **do**

7   // Calculate the new weight of objects by Eq. (8.1)

8   *weight_of_objects* := *null*;

9   **for** *o* = 1 **to** *number_of_objects* **do**

10   *wrong_classification* := classified_object(*o, meta_ensemble*);

11   **if** *wrong_classification* == *0* **then**

12   *weight_of_objects[o]* := 1;

13   **endif**

14   **else**

15   *weight_of_objects[o]* = *wrong_classification · n*;

16   **endif**

17   **endfor**

18   // Classifier construction

19   *data_set_classifier* :=choose_objects(*data_set, weight_of_objects*); // by Eq. (8.2)

20   *new_classifier* := build_tree_using_ACDT(*data_set_classifier, pheromone*);

21   **if** *new_classifier* **is_of_higher_quality_than** *best_classifier* **then**

22   *best_classifier* := *new_classifier*;

23   **endif**

24   **endfor**

25   update_pheromone_trail(*best_classifier, pheromone*);

26   *meta_ensemble*.add(*best_classifier*);

27 **endfor**

28 *result* := *meta-ensemble*;

---

self-adaptive ant colony decision forest (saACDF). Such a solution was aimed at using self-adaptive capabilities for adjusting the object weights on a current basis—while learning the set of classifiers. In that case, dynamic pseudo-samples are chosen according to the previously obtained classification quality—unlike the aACDF approach, where the algorithm itself determines the relative weights of objects.

Self-adaptation consists in either increasing or decreasing the probability of choosing the individual objects to a sample (incorrect classification of objects to a specific class carried out by simple classifiers). At the beginning, the probability of being chosen is the same for each object, and equals $\frac{1}{n}$. For the next population, that probability depends on changing the weight for each object in the training set. The weight is set in accordance with the results for the previous population created by the best ant agent. This induces a greater probability of choosing trouble-causing objects to the next pseudo-sample. The above is the difference with respect to the ACDF approach, in which two cases are examined: the correct and the wrong classification. The weight of object x_i is calculated according to the formula:

Algorithm 10: Self-adaptive ACDF algorithm

---

1 pheromone_trail_initialization(*pheromone*);
2 *weight_of_objects[1..n]* := $\frac{1}{n}$;
3 *meta_ensemble* := *null*;
4 **for** $j = 1$ **to** *number_of_classifiers* **do**
5   *best_classifier* := null;
6   **for** $a = 1$ **to** *ants_per_colony* **do**
7   // Calculate the new weight of objects by Eq. (8.3)
8   *weight_of_objects* := *null*;
9   **for** $o = 1$ **to** *number_of_objects* **do**
10   *wrong_classification* := classified_object(*o, meta_ensemble*);
11   **if** *wrong_classification* $== 0$ **then**
12   *weight_of_objects[o]* := 1;
13   **endif**
14   **else**
15   *weight_of_objects[o]* $= wrong\_classification \cdot n$;
16   **endif**
17   **endfor**
18   // Classifier construction
19   *data_set_classifier* :=choose_objects(*data_set, weight_of_objects*); // by Eq. (8.2)
20   *new_classifier* := build_tree_using_ACDT(*data_set_classifier, pheromone*);
21   **if** *new_classifier* **is_of_higher_quality_than** *best_classifier* **then**
22   *best_classifier* := *new_classifier*;
23   **endif**
24   **endfor**
25   update_pheromone_trail(*best_classifier, pheromone*);
26   *meta_ensemble*.add(*best_classifier*);
27 **endfor**
28 *result* := *meta-ensemble*;

$$we(x_i) = \begin{cases} 1, & \text{if } w = 0 \\ w \cdot n & \text{otherwise,} \end{cases} \qquad (8.3)$$

where w is the number of decision trees which have incorrectly (wrongly) classified the object x_i. Then the probability of choosing the object is calculated according to the formula (8.2).

Self-adaptive ACDF is a sequential learning schema, in which the most important factor is the current form of the ensemble. It is not classical sequential training, like that seen in boosting, where the probability of object selection is determined by the homogeneous basic classifiers present in the ensemble (one step back). In this case, the classification history is not significant (line 4 and 6 in pseudo-code—Algorithm 10).

As can be noted, in this particular case the weight of every single object in the learning set depends not only on the fact whether it has been wrongly classified by the ensemble of classifiers or not, but also on the number of classifiers (currently included in the ensemble) that have wrongly classified that object. Such a solution fundamentally distinguishes saACDF from aACDF and eliminates the problem that could be seen in case of tuning the parameter λ. The experiments with the described ensemble methods were presented in Sect. 8.4.2.

8.3 Ant Colony Decision Forest Based on Boosting

ACDF-Boost is an approach that combines the ACDF algorithm with the ideas that were first applied in AdaBoost. For consecutive populations, pseudo-samples are created through weighted sampling, where each object is assigned the weight calculated according to the formula (8.4) and then normalized (8.2). The ACDF-Boost algorithm differs from the classical approach in its way of operation. The ensemble of classifiers is created by trees that have been built with the ACDT algorithm—after evaluating the results obtained by one population of ants, a single tree is chosen. Similarly to the random forest, decision trees are constructed using a limited number of attributes.

In this case, ant agents construct the tree based on previously preserved pheromone trails over simultaneously changing pseudo-samples that have been determined by the errors in classification of objects. This approach is enabled by a distinct form of distributed tree construction. The weight of object x_i at time t is calculated as follows:

$$we(x_i, t) = \begin{cases} we(x_i, t-1), & \text{if object } x_i \text{ is well classified} \\ we(x_i, t-1) \cdot e^{\left(\frac{1}{2} \cdot \ln(ac(x_i))\right)} & \text{otherwise,} \end{cases} \quad (8.4)$$

whereas $ac(x_i)$ is defined by the votes of single classifiers for object x_i (using simple voting):

$$ac(x_i) = \frac{1 - sv(x_i)}{sv(x_i)}, \quad (8.5)$$

Here $sv(x_i)$ denotes the relationship between the number of trees in the ensemble that have correctly classified the object x_i and the current number of (previously created) trees in the analyzed ensemble. In this approach, weights of the objects are calculated irrespectively of the previous classification results.

Algorithm 11: The ACDF-Boost algorithm

---

1 initialization_pheromone_trail(*pheromone*);

2 *weight_of_objects[1..n]* := $\frac{1}{n}$;

3 *meta_ensemble* := *null*;

4 *weight_of_objects* := *null*;

5 **for** *j* = 1 **to** *number_of_classifiers* **do**

6   *best_classifier* := null;

7   **for** *a* = 1 **to** *ants_per_colony* **do**

8   // Calculate new weights of objects using Eq. (8.4)

9   **for** *o* = 1 **to** *number_of_objects* **do**

10   classifies_object(*o, meta_ensemble*);

11   **if** *if object o is not classified well* **then**

12   *weight_of_objects[o]* := *weight_of_objects[o]* · *w*; // where *w* is calculated from Eq. (8.4)

13   **endif**

14   **endfor**

15   // Classifier construction

16   *data_set_classifier* :=choose_objects(*data_set, weight_of_objects*);

17   *new_classifier* := build_tree_using_ACDT(*data_set_classifier, pheromone*);

18   **if** *new_classifier* **is_of_higher_quality_than** *best_classifier* **then**

19   *best_classifier* := *new_classifier*;

20   **endif**

21   **endfor**

22   update_pheromone_trail(*best_classifier, pheromone*);

23   *meta_ensemble*.add(*best_classifier*);

24 **endfor**

25 *result* := *meta-ensemble*;

In opposition to saACDF, ACDF-Boost is based on the classification history, not only on the previously obtained results. As the term "sequential" determines, we learn from the results of consecutive homogeneous classifiers. The probability of object selection to the pseudo-sample depends on the given iteration of the ensemble (see lines 2 and 8 of pseudo-code—Algorithm 11).

The ACDF-Boost approach allows for simultaneous application of proportionality by setting support for badly classified objects. In other words, probability of object selection for the next pseudo-sample (of course, also multiple selection, because draw with return is still used) is higher if a greater number of single classifiers in the ensemble classify the given object wrongly. Similarly as in case of other algorithms described in this chapter, the research experiments are shown in Sect. 8.4.2.

8.4 Computational Experiments

To verify the ACDF models described in this chapter, three experiments were conducted, in which, for example, comparisons of the described algorithms with classical ensemble methods known from the literature were conducted. The experiments were subjected to additional statistical tests.

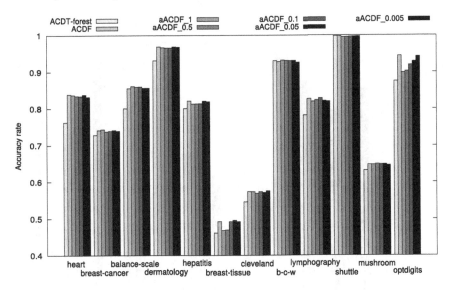

Fig. 8.1 Accuracy rate versus values of parameter λ

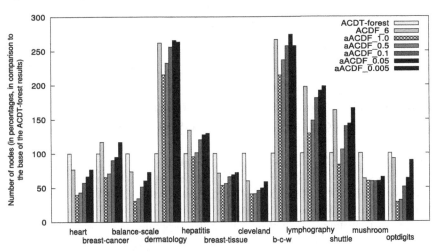

Fig. 8.2 Number of nodes (in percents, compared to the results for base ACDT-forest)

To maintain the continuity of research, the same data sets were used each time (Table 7.2), with the same way of testing the results. Additionally, the number of iterations, the population size and the ensemble of classifiers size were all compatible with the assumptions set forth in Sect. 7.3.

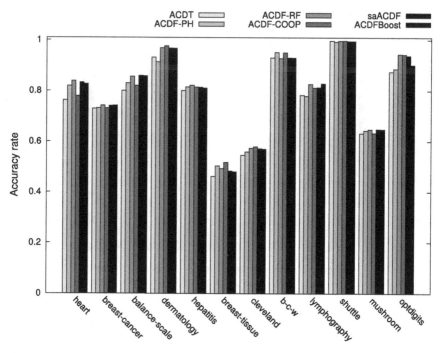

Fig. 8.3 Accuracy rate

8.4.1 Examination of the Adaptation Parameter in the Adaptive Ant Colony Decision Forest Algorithm

The proposed version of adaptive ACDF should, above all, be compared with its predecessors: ACDT-forest and ACDF-RF (based on random forest) for different values of parameter λ. The values of λ examined in this chapter were established arbitrarily as 1.0, 0.5, 0.1, 0.05 and 0.005.

The experimental results confirm that adaptively applied sample subsets allow for obtaining significantly better results (Table 8.1). For the majority of data sets, in most cases it was possible to construct decision forests consisting of smaller decision tress (with a reduced number of nodes) compared to the ACDT-forest and the ACDF-RF algorithms, as can be seen in Fig. 8.2. Reduction in decision tree growth depended on the value of λ. With the growing value of λ, smaller decision trees and ensembles were created. The influence of λ on the classification quality can be also observed (see Fig. 8.1). Namely, use of the values of λ greater than 1.0 causes deterioration in the classification quality.

Table 8.1 Comparative study—examination of λ parameter values

Data set	ACDT—forest		ACDF-RF		aACDF ($\lambda = 1.0$)	
	Acc.	#n	Acc.	#n	Acc.	#n
heart	0.7628	253.4	**0.8391**	194.8	0.8372	**101.39**
breast-cancer	0.7284	82.1	0.7414	96.0	**0.7434**	**53.80**
balance-scale	0.8003	1198.8	0.8559	884.3	**0.8613**	**365.70**
dermatology	0.9314	**113.7**	**0.9690**	297.7	0.9667	244.79
hepatitis	0.8005	43.4	**0.8210**	58.4	0.8125	**41.46**
breast-tissue	0.4611	205.1	0.4919	146.0	0.4682	**109.37**
cleveland	0.5456	338.7	0.5737	202.9	0.5732	**137.71**
bcw	0.9306	**103.2**	0.9280	275.1	**0.9319**	221.03
lymphography	0.7821	**161.6**	0.8265	318.5	0.8193	208.29
shuttle	**0.9975**	1508	0.9972	2458	0.9945	**1253**
mushroom	0.6313	1604	0.6475	1007	0.6469	950
optdigits	0.8751	4273	**0.9442**	3942	0.8985	**1203**

Data set	aACDF							
	$\lambda = 0.5$		$\lambda = 0.1$		$\lambda = 0.05$		$\lambda = 0.005$	
	Acc.	#n	Acc.	#n	Acc.	#n	Acc.	#n
heart	0.8348	109.65	0.8344	145.99	0.8385	169.10	0.8323	193.96
breast-cancer	0.7374	58.05	0.7393	74.12	0.7411	78.04	0.7389	95.77
balance-scale	0.8597	412.33	0.8601	617.18	0.8565	722.26	0.8565	870.88
dermatology	0.9661	264.29	0.9661	291.10	**0.9690**	302.16	0.9681	299.41
hepatitis	0.8130	43.97	0.8134	52.12	0.8203	55.42	0.8188	56.26
breast-tissue	0.4696	115.22	0.4922	135.29	**0.4954**	140.85	0.4922	147.77
cleveland	0.5687	138.89	0.5733	155.34	0.5710	167.29	**0.5751**	197.87
bcw	0.9309	244.23	0.9306	265.52	0.9310	282.60	0.9262	265.40
lymphography	0.8231	238.93	**0.8282**	291.70	0.8211	309.53	0.8197	319.77
shuttle	0.9951	1585	0.9960	2104	0.9962	2162	0.9970	2500
mushroom	**0.6492**	944	0.6485	**940**	0.6488	947	0.6458	1043
optdigits	0.9024	1331	0.9193	2171	0.9290	2707	0.9422	3828

Abbreviations: *Acc.* accuracy rate; *#n* number of nodes

8.4.2 Comparison with Other Algorithms

The results of experiments confirm that use of self-adaptive mechanisms in construction of data subsets allows for obtaining better results (see Table 8.2 and Fig. 8.3). For the majority of analyzed data sets, the use of such a method enables construction of an ensemble of classifiers with considerably smaller decision trees than those obtained with the previously analyzed ACDT and ACDF approaches (Chap. 7). This can be observed when analyzing the results presented in Fig. 8.4. Smaller decision

Table 8.2 Comparative study—accuracy rate and number of nodes (standard deviations in parentheses)

Data set	ACDT—forest		ACDF-PH		ACDF-RF		ACDF-COOP		aACDF ($\lambda = 1.0$)		saACDF		ACDF-Boost	
	Acc.	#n	Acc.	#n	Acc.	#n	Acc.	#n	Acc.	#n	Acc.	#n	Acc.	#n
heart	0.7628	253.4	0.8191	155.1	**0.8391**	194.8	0.7781	257.8	0.8372	101.39	0.8324	158.67	0.8270	110.5
breast-cancer	0.7284	82.1	0.7313	92.7	0.7414	96.0	0.7308	122.9	**0.7434**	53.80	0.7398	76.1	0.7415	59.8
balance-scale	0.8003	1198.8	0.8299	426.9	0.8559	884.3	0.8202	1135.7	**0.8613**	365.70	0.8591	716.4	0.8581	417.2
dermatology	0.9314	113.7	0.9135	242.8	0.9690	297.7	**0.9765**	272.1	0.9667	244.79	0.9673	261.6	0.9672	263.9
hepatitis	0.8005	43.4	0.8152	77.8	**0.8210**	58.4	0.8145	79.9	0.8125	41.46	0.8137	44.0	0.8106	43.9
breast-tissue	0.4611	205.1	0.5029	151.7	0.4919	146.0	**0.5167**	198.5	0.4682	109.37	0.4826	140.8	0.4795	114.8
cleveland	0.5456	338.7	0.5590	165.7	0.5737	202.9	**0.5795**	291.4	0.5732	137.71	0.5714	178.3	0.5695	138.1
bcw	0.9306	103.2	**0.9519**	146.1	0.9280	275.1	0.9505	127.1	0.9319	221.03	0.9303	190.2	0.9300	232.97
lymphography	0.7821	161.6	0.7782	115.5	0.8265	318.5	0.8122	167.5	0.8193	208.29	0.8131	291.6	**0.8283**	184.2
shuttle	0.9975	1508	0.9934	7992	0.9972	2458	**0.9976**	1510	0.9945	1253	0.9955	1323	0.9952	1589
mushroom	0.6313	1604	0.6420	579	0.6475	1007	0.6331	1617	0.6469	950	**0.6481**	1002	0.6470	956
optdigits	0.8751	4273	0.8862	4725	**0.9442**	3942	0.9429	4235	0.8985	1203	0.9381	3552	0.9011	1351

Abbrevations: *Acc.* accuracy rate; *#n* number of nodes

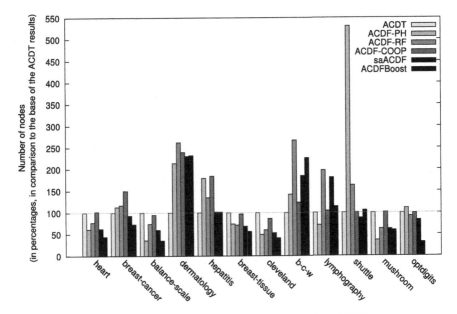

Fig. 8.4 Number of nodes (in percents, compared to the results for base ACDT)

trees in the decision forest mean that the ensemble of classifiers takes up much less space and allows faster classification.

In some cases, the classification accuracy is improved, even though for a vast number of objects it is on the same level as for ACDF-RF and other original approaches without self-adaptation. We should stress here that the quality of an ensemble of classifiers created with aACDF, saACDF or ACDF-Boost is better. The substantially higher quality can be observed in the diagram presented in Fig. 8.5, where the relationship between decision tree growth and classification accuracy depending on the applied algorithms is also highlighted. It is noteworthy that aACDF and ACDF-Boost usually not only yield good classification results, but also create smaller decision trees in the ensemble of classifiers. The hypotheses developed based on ACDF-Boost usually yield the best classification accuracy for data sets characterized by a small number of decision classes (2–3). In turn, better classification accuracy is provided by the saACDF approach for data sets characterized by a large number of decision classes and a big number of attributes. The ACDF-COOP approach was proven good in case of small data sets characterized by a small number of attributes, as well as a small number of decision classes.

The ACDF approach is also compared with the classical AdaBoost.M1, AdaBoost.M1 with the CART classifier (implemented in the WEKA system [5]), the random forest (also implemented in WEKA), and the support vector machine (SVM) [6]. The results obtained (see Table 8.3) demonstrate that our proposal performs better in most cases. The greatest differences in classification accuracy (and

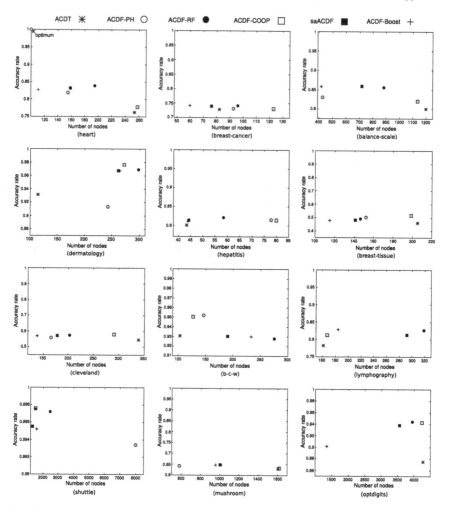

Fig. 8.5 Relationship between decision tree growth and classification accuracy depending on the applied algorithm

the number of nodes for AdaBoost.M1) can be observed when the data sets contain a significant number of decision classes.

8.4.3 Statistical Analysis

Just like in case of the ACDF algorithms described in Chap. 7, also the adaptive approach was analyzed statistically. The results can bee seen in Tables 8.4, 8.5, 8.6, 8.7, 8.8 and 8.9. For the statistical checks, values of the λ parameter, were used,

Table 8.3 Comparison of ACDF-Boost with AdaBoost.M1

Data set	ACDF-Boost		Ada-Boost.M1	AdaBoost.M1 (CART)		random forest	SVM
	Acc.	Nodes	Acc.	Acc.	Nodes	Acc.	Acc.
heart	**0.8270**	110.5	0.8111	0.5555	117	0.6962	0.7928
breast-cancer	**0.7415**	59.8	0.7290	0.6531	493	0.6591	0.7148
balance-scale	0.8581	417.2	0.7068	0.7808	1831	0.7598	**0.9167**
dermatology	**0.9672**	263.9	0.5027	0.9542	257	0.9123	0.9510
hepatitis	0.8106	43.9	**0.8316**	0.7416	61	0.8049	0.7937
breast-tissue	**0.4795**	114.8	0.2026	0.2879	94	0.2319	0.2298
cleveland	**0.5695**	138.1	0.5148	0.4527	409	0.4922	0.5442
bcw	0.9300	233.0	0.9356	**0.9484**	365	0.9356	0.9442
lymphography	0.8283	184.2	0.8367	**0.8456**	275	0.8367	0.8367
shuttle	**0.9952**	1589	0.8519	0.9912	1337	0.9941	0.9931
mushroom	**0.6470**	956	0.5029	0.5111	18,170	0.5011	0.6177
optdigits	**0.9011**	1351	0.1898	0.8987	3190	0.7601	0.8999

Abbreviations: *Acc.* accuracy rate; *nodes* number of nodes

Table 8.4 Differences between ACDF algorithms (critical differences in boldface)—accuracy rate

Accuracy

	ACDT forest	ACDF PH	ACDF RF	ACDF COOP	aACDF λ = 1.0	saACDF	ACDF Boost	AdaBoost M1	AdaBoost M1 (CART)	Random forests	SVM
ACDT-for.	–	1.9167	**4.2500**	**3.6667**	**3.0000**	**3.4167**	**2.8333**	−0.3750	−0.3333	−0.9583	0.9167
ACDF-PH	−1.9167	–	**2.3333**	1.7500	1.0833	1.5000	0.9167	**−2.2917**	−2.2500	**−2.8750**	−1.0000
ACDF-RF	**−4.2500**	**−2.3333**	–	−0.5833	−1.2500	−0.8333	−1.4167	**−4.6250**	**−4.5833**	**−5.2083**	**−3.3333**
ACDF-COOP	**−3.6667**	−1.7500	0.5833	–	−0.6667	−0.2500	−0.8333	**−4.0417**	**−4.0000**	**−4.6250**	**−2.7500**
aACDF	**−3.0000**	−1.0833	1.2500	0.6667	–	0.4167	−0.1667	**−3.3750**	**−3.3333**	**−3.9583**	−2.0833
saACDF	**−3.4167**	−1.5000	0.8333	0.2500	−0.4167	–	−0.5833	**−3.7917**	**−3.7500**	**−4.3750**	**−2.5000**
ACDF-Boost	**−2.8333**	−0.9167	1.4167	0.8333	0.1667	0.5833	–	**−3.2083**	**−3.1667**	**−3.7917**	−1.9167
AdaBoost.M1	0.3750	**2.2917**	**4.6250**	**4.0417**	**3.3750**	**3.7917**	**3.2083**	–	0.0417	−0.5833	1.2917
AdaB.(CART)	0.3333	2.2500	**4.5833**	**4.0000**	**3.3333**	**3.7500**	**3.1667**	−0.0417	–	−0.6250	1.2500
RF	0.9583	**2.8750**	**5.2083**	**4.6250**	**3.9583**	**4.3750**	**3.7917**	0.5833	0.6250	–	1.8750
SVM	−0.9167	1.0000	**3.3333**	**2.7500**	2.0833	**2.5000**	1.9167	−1.2917	−1.2500	−1.8750	–

Table 8.5 Friedman test results and mean ranks (the best rank in boldface)—examination of the λ parameter values

Accuracy

	Values
N	30
Chi-square	5.755274
Degrees of freedom	4
p value is less than	0.2182
5% critical difference	1.266479
Mean ranks	
aACDF $\lambda = 1.0$	3.3333
aACDF $\lambda = 0.5$	3.5417
aACDF $\lambda = 0.1$	2.9167
aACDF $\lambda = 0.05$	**2.1250**
aACDF $\lambda = 0.005$	3.0833

Table 8.6 Differences between ACDF algorithms (critical differences in boldface)—examination of the λ parameter values

Accuracy

	aACDF $\lambda = 1.0$	aACDF $\lambda = 0.5$	aACDF $\lambda = 0.1$	aACDF $\lambda = 0.05$	aACDF $\lambda = 0.005$
aACDF $\lambda = 1.0$	–	−0.2083	0.4167	1.2083	0.2500
aACDF $\lambda = 0.5$	0.2083	–	0.6250	**1.4167**	0.4583
aACDF $\lambda = 0.1$	−0.4167	−0.6250	–	0.7917	−0.1667
aACDF $\lambda = 0.05$	−1.2083	**−1.4167**	−0.7917	–	−0.9583
aACDF $\lambda = 0.005$	−0.2500	−0.4583	0.1667	0.9583	–

along with the non-parametric Friedman test for comparing different algorithms. In Tables 8.4, 8.6 and 8.9, bold fonts indicate the values meeting the criterion of 5% critical difference. Tables 8.5, 8.7 and 8.8 present all values determined during the statistical analyses, including mean ranks for the samples.

As may be seen in Table 8.5 in the case of analyzing the values of parameter λ, the best rank was obtained by the algorithm for $\lambda = 0.05$. However, from the earlier Table 8.1 we could have concluded that a better rank would be achieved for $\lambda = 1.0$ (which eventually got the second rank). It should be noted that only in a single case there is a critical difference between the obtained results—namely, in case of $\lambda = 0.05$ compared to $\lambda = 0.5$ (see Table 8.6).

Table 8.7 Friedman test results and mean ranks (the best rank in boldface)—accuracy rate

Accuracy	
	Values
N	30
Chi-square	38.622391
Degrees of freedom	10
p value less than	0.0001
5% critical difference	2.280454
Mean ranks	
ACDT-forest	7.6667
ACDF-PH	5.7500
ACDF-RF	**3.4167**
ACDF-COOP	4.0000
aACDF	4.6667
saACDF	4.2500
ACDF-Boost	4.8333
AdaBoost.M1	8.0417
AdaBoost.M1 (CART)	8.0000
Random forests	8.6250
SVM	6.7500

Table 8.8 Friedman test results and mean ranks (the best rank in boldface)—number of nodes

Number of nodes	
	Values
N	30
Chi-square	26.250000
Degrees of freedom	7
p value is less than	0.0005
5% critical difference	1.724257
Mean ranks	
ACDT-forest	4.6667
ACDF-PH	4.2500
ACDF-RF	6.0000
ACDF-COOP	6.0000
aACDF	**2.0000**
saACDF	4.1667
ACDF-Boost	3.4167
AdaBoost.M1 (CART)	5.5000

Table 8.9 Differences between ACDF algorithms (critical differences in boldface)—number of nodes

Number of nodes

	ACDT forest	ACDF PH	ACDF RF	ACDF COOP	aACDF $\lambda = 1.0$	saACDF	ACDF Boost	AdaBoost.M1 (CART)
ACDT-for.	–	0.4167	–1.3333	–1.3333	**2.6667**	0.5000	1.2500	–0.8333
ACDF-PH	–0.4167	–	**–1.7500**	**–1.7500**	**2.2500**	0.0833	0.8333	–1.2500
ACDF-RF	1.3333	**1.7500**	–	0.0000	**4.0000**	**1.8333**	**2.5833**	0.5000
ACDF-COOP	1.3333	**1.7500**	0.0000	–	**4.0000**	**1.8333**	**2.5833**	0.5000
aACDF	**–2.6667**	**–2.2500**	**–4.0000**	**–4.0000**	–	**–2.1667**	–1.4167	**–3.5000**
saACDF	–0.5000	–0.0833	**–1.8333**	**–1.8333**	**2.1667**	–	0.7500	–1.3333
ACDF-Boost	–1.2500	–0.8333	**–2.5833**	**–2.5833**	1.4167	–0.7500	–	**–2.0833**
AdaB.(CART)	0.8333	1.2500	–0.5000	–0.5000	**3.5000**	1.3333	**2.0833**	–

Table 8.10 Characteristics of analyzed e-mail data set

Data set	No. of instances	No. of classes (e-mail folders)	Number of value of attributes					
			from	word1	word2	word3	cc	length
beck-s	1971	101	390	527	670	549	2	1331
farmer-d	3672	25	412	827	985	864	2	1679
kaminski-v	4477	41	821	1231	1304	1058	2	2461
kitchen-l	4015	46	597	1170	1207	996	2	2138
lokay-m	2493	11	295	842	955	863	2	1654
sanders-r	1188	30	272	442	485	423	2	1033
williams-w3	2769	18	196	523	597	540	2	1056

Table 8.11 Comparison of ensembles of classifiers in terms of classification accuracy

Data set	AdaBoost		Dagging		Bagging CART	Random Forest	aACDF
	CART	RandomTree	CART	RandomTree			
beck-s	0.384	0.184	0.411	0.446	0.416	0.481	**0.517**
farmer-d	0.620	0.658	0.441	0.548	0.587	0.670	**0.775**
kaminski-v	0.271	0.105	0.123	0.385	0.250	0.349	**0.657**
kitchen-l	0.292	–	0.178	0.471	0.321	0.264	**0.583**
lokay-m	0.685	0.638	0.465	0.523	0.723	0.461	**0.846**
sanders-r	0.480	0.758	0.354	0.434	0.457	0.649	**0.759**
williams-w3	0.887	0.914	0.714	0.730	0.859	0.819	**0.944**

In case of statistical analysis, it was useful to compare different versions of ACDF with different ensembles of classifiers. The comparison was conducted for both the classification accuracy and the ensemble size (counted as the number of nodes for all trees in the forest). For this reason, the ensemble size analysis was only conducted for selected approaches.

As can bee seen in Table 7.4, the best algorithm (with respect to the classification accuracy) was ACDF-RF, which achieved the best rank. A critical difference (see Table 7.5) was achieved compared to the remaining ensembles not related to ACO. As to the comparison with adaptive versions of ACDF (described in this chapter), ACDF-RF was better, but not critically better. Quite different results (in terms of adaptability) were obtained from the analysis concerning the number of nodes. Here the best algorithms seemed to be adaptive ACDF variants, with the highest ranks achieved by, respectively: aACDF, ACDF-Boost and saACDF. A slightly worse result was achieved by ACDF-PH, while ACDF-RF (along with ACDF-COOP, which was not a surprise) obtained the worst ranks of all.

Experimental results confirm that application of ant colony algorithms in construction of classifiers allows for better analysis of the whole solution space, resulting in object classification improvement, and at the same time decreasing the classifier size.

8.5 Example of a Practical Application

The ACDF algorithm was successfully applied to a real-world data set related to the e-mail foldering problem [1]. In [4], aACDF was used (as described in Sect. 8.1) to predict the labels of e-mail messages originating from the Enron e-mail data set. The e-mail foldering problem, along with the mentioned data set, was described in detail in Part I of this book (Sect. 5.2).

Probierz et al. proposed application of the original version of aACDF (without modifications) to the problem of analyzing raw data included in e-mail messages, which represented in some sense information noise. Additionally, the e-mail messages were preprocessed and presented as decision tables with a very large number of decision classes (Table 8.10). Then, an ACO based ensemble of classifiers was compared with other ensemble methods. The results obtained by the authors are presented in Table 8.11.

It can be seen that when working on raw, real-world data with a large number of decision classes (having irregular distribution), the adaptive ant colony decision forest allows for obtaining far better results than other approaches. Both boosting and bagging, as well as the random forest, yielded definitely worse predictions of the e-mail folders from the test set.

8.6 Conclusions

An ensemble method with ant agents leads to certain effects known as: autocatalysis, positive feedback method in pheromone representations, and finally self-organization in the ant colony behavior. Some correlation between these mechanisms and innovation in the area of data mining can be noticed. This new form of innovative methodology leads to a collective intelligence spreading within this artificial organism. The above innovation should be summarized as a new form of simulation, concentrating on the "life" of the ACDF system. The wisdom of artificial ants in ACDF depends on the number of attributes and their values, and in consequence on the number of nodes included in decision trees. The number of nodes and the strength of the connections expressed by pheromone values arise from the aggregation of the ACO approach with the CART algorithm.

In this approach, a very important issue is understanding the power of information exchange via pheromone-collaborative efforts, and the use of different channels in this structure, called heterarchy [3, 7]. The introduction of such an innovation in the data mining tasks for the purpose of reinforcing ant learning in ACDF (collective intelligence) is intended to provide a comprehensive approach to difficult, challenging and significant classification problems.

The use of ant colony algorithms in decision tree construction allows us to construct a variety of alternative decision trees, representing different local optima. A lot of decision trees with different local optima allow for considering a broader range of solutions by an ensemble of classifiers. The classification becomes more stable. In case of deterministic algorithms, each decision tree is constructed in the same way.

In ACDF, the weights of objects in pseudo-samples have been demonstrated to yield better results in a comparative study with other approaches. The advantages of aACDF and ACDF-Boost over other approaches may be expected to be significantly higher in an application yielding smaller decision trees with considerably better classification accuracy. The results confirm that the proposed method allows for building much smaller classifiers while maintaining high classification accuracy. This means that self-adaptation allows for finding the individual classifiers more relevant to the ensemble of classifiers. The subsequent classifiers are built according to the current ensemble of classifiers.

The process of accommodating pseudo-samples in the current population of ant agents allowed for constructing decision trees which outperformed the previous results. This is due to a greater attention being paid to incorrectly classified objects.

It should be emphasized that, in contrast to other ensemble methods, ACDF-related approaches all the time have information about the structure of previously built decision trees, which is memorized in the form of a pheromone trail. Such a feedback supplements the information about the classification of some concrete objects with additional information on the construction of previous classifiers. In this case, the results achieved at a given time by the ensemble of classifiers is given along with the way of constructing those classifiers.

References

1. R. Bekkerman, A. McCallum, G. Huang, Automatic categorization of email into folders: benchmark experiments on enron and sri corpora. *Center for Intelligent Information Retrieval, Technical Report IR* (2004)
2. U. Boryczka, J. Kozak, On-the-go adaptability in the new ant colony decision forest approach, in *Asian Conference on Intelligent Information and Database Systems* (Springer International Publishing, 2014), pp. 157–166
3. U. Boryczka, J. Kozak, R. Skinderowicz, Heterarchy in constructing decision trees—parallel acdt. T. Comp. Collect. Intell. **10**, 177–192 (2013)
4. U. Boryczka, B. Probierz, J. Kozak. Adaptive ant colony decision forest in automatic categorization of emails, in *Asian Conference on Intelligent Information and Database Systems* (Springer, 2015), pp. 451–461
5. R.R. Bouckaert, E. Frank, M. Hall, R. Kirkby, P. Reutemann, A. Seewald, D. Scuse, Weka manual for version 3-7-10 (2013)
6. C.C. Chang, C.J. Lin, LIBSVM: a library for support vector machines. ACM Trans. Intell. Syst. Technol. 2:27:1–27:27, 2:1–27 (2011)
7. J. Dreo, P. Siarry, Continuous interacting ant colony algorithm based on dense heterarchy. Future Gener. Comput. Syst. 841–856 (2004)
8. Y. Freund, R.E. Schapire, Experiments with a new boosting algorithm, in *International Conference on Machine Learning* (1996), pp. 148–156
9. M. Kearns, Thoughts on hypothesis boosting. Project for Ron Rivest's machine learning course at MIT (1988)
10. J. Kozak, U. Boryczka, Multiple boosting in the ant colony decision forest meta-classifier. Knowl. Based Syst. **75**, 141–151 (2015)
11. R.E. Schapire, The strength of weak learnability. Mach. Learn. **5**, 197–227 (1990)

Chapter 9
Summary

9.1 Final Remarks

This book is oriented at the use of ant colony optimization as an algorithm and method for constructing decision trees and ensemble classifiers. When preparing the text, we tried to stick to a uniform approach and presentation of results. This was not always possible—especially in case of analysis of real-world applications, and in case of single classifiers—in contrast to ensemble classifiers. Nevertheless, we tried to keep the conditions of experiments and their descriptions as uniform as possible throughout the book.

Writing an introduction to this book was a two-fold, extremely difficult challenge. On the one hand, it had to be compact, and on the other hand—complete. As collective intelligence algorithms and machine learning algorithms are considered to represent two different areas of knowledge, the first chapter not only introduces detailed solutions, but also contains descriptions of classical algorithms and simplified examples that allow the reader to understand the essence of the book. Different examples can be found in Chap. 3, where we tried to show examples of learning decision trees using ant colony optimization.

Both parts of the book are organized in essentially the same way. They begin with an introduction to the existing problems: first, evolutionary computing techniques in data mining are presented, and then evolutionary computing techniques in ensemble learning—all followed by a detailed literature review. After this, the core concept of combining ant colony optimization with decision tree or ensemble methods is presented, and finally examples of practical applications are given.

In case of all computational experiments, statistical analysis was performed—with the same methodology used in every chapter. The only algorithm presented in a greater depth is ACDT. To compare it with classical algorithms, additional statistical tests were introduced—while maintaining those performed in the whole book.

The general purpose of this book was to describe the possibilities of using collective intelligence (here represented by ant colony optimization) in the machine

© Springer International Publishing AG, part of Springer Nature 2019   157
J. Kozak, *Decision Tree and Ensemble Learning Based on Ant Colony Optimization*, Studies in Computational Intelligence 781,
https://doi.org/10.1007/978-3-319-93752-6_9

learning process (here—decision trees and ensemble methods). To fully attain this goal, it was necessary to present the existing solutions while also trying to explain why those algorithms bring good results.

For this purpose, additional analyses of the impact on the achieved results of aspects known from ant colony optimization—here, mostly aspects concerning the analysis of pheromone maps—were carried out. This allowed us to show the adaptability of ant colony optimization, and in general the possibility to search a large solution space (producing larger sets of local optima—potentially similar solutions) as the aspect with the greatest influence on good results of the algorithms (as presented in Chaps. 3, 4, 7 and 8). The above is the main difference compared to the classical approaches commonly used for constructing decision trees—the introduced methods bring the algorithm closer to ensemble methods.

9.2 Future Directions

The introduction of an ant colony decision tree and an ant colony decision forest seems to summarize the theory of decision trees and ensemble methods based on ant colony optimization. In fact, it opens new possibilities of their use. A few open problems are also presented in this section.

On the basis of the conclusions presented above, it may be easily seen that decision trees based on an ant colony can be applied to different problems, for which the adaptability to dynamic data can be a crucial aspect. The whole, completely new concept is the construction of a classifier on the basis of a single learning set which is not significantly modified and allows for using the classifier without the necessity of its update. Something completely different is the construction of a classifier adapted to a training set which is constantly changed as new objects are added to it. Such data change can be performed continuously or with the use of data—which is most often called stream data.

As may be supposed, it is possible to design an algorithm that would allow for exploration of data streams based on cooperation of agents (for example, via the pheromone known from ant colony optimization) and some additional knowledge (data packages) acquired during the algorithm run. By a data stream we understand a (potentially infinite) data sequence, in which data items are passed to the system as packages or as a continuous stream. However, there is no possibility of storing historical data in the memory; moreover, the algorithm should quickly adapt the classifier to the new data distribution. Development of an algorithm based on collective intelligence could (by some sort of natural self-adaptation) minimize the mentioned problem resulting from the possibility of data distribution change (so-called concept drift). In case of positive results, development of a similar algorithm will open the road to using ant colony optimization in big data analysis, which is very important nowadays.

There are still a lot of new possibilities of ensemble learning based on ant colony optimization. It can be noted that the algorithms described in this book which are based on ant colony optimization and used for ensemble learning have not been examined for weighted voting. The current versions work in a simple way—similar to bagging, where every single classifier has a vote with the same weight. However, it should be checked what results could be obtained by the algorithm enhanced with weighted voting. It can be used similarly to boosting, but can be also developed on the basis of the possibilities introduced by ant colony optimization (for example, it can be based on the pheromone trail value acquired from every single classifier).

This should also be checked for other solutions related to the construction of a single classifier. Nowadays it could be said (roughly speaking) that the ACDF algorithm is based on the actual shape of the ensemble classifier in case of pseudo-sample construction; however, selection of the features, or node division, takes place locally. Such solutions are justified, but one can also introduce more cooperation between the ant agents in case of the remaining calculations, too.

Printed in the United States
By Bookmasters